ELEMENTARY ORGANIC CHEMISTRY

To
V. and my Parents

ELEMENTARY ORGANIC CHEMISTRY

D. E. F. ARMSTEAD, B.Sc. (Hons), A.R.I.C.

LONDON
BUTTERWORTHS

ENGLAND: BUTTERWORTH & CO. (PUBLISHERS) LTD.
LONDON: 88 Kingsway, W.C.2

AUSTRALIA: BUTTERWORTH & CO. (AUSTRALIA) LTD.
SYDNEY: 20 Loftus Street
MELBOURNE: 343 Little Collins Street
BRISBANE: 240 Queen Street

CANADA: BUTTERWORTH & CO. (CANADA) LTD.
TORONTO: 14 Curity Avenue, 374

NEW ZEALAND: BUTTERWORTH & CO. (NEW ZEALAND) LTD.
WELLINGTON: 49/51 Ballance Street
AUCKLAND: 35 High Street

SOUTH AFRICA: BUTTERWORTH & CO. (SOUTH AFRICA) LTD.
DURBAN: 33/35 Beach Grove

Suggested U.D.C. Number: 547
Standard Book Number: 408 460601

Printed in Great Britain by Page Bros. (Norwich) Ltd., Norwich

PREFACE

THIS book is designed to meet the needs of students preparing for the City and Guilds General Course in Science and the General Certificate of Education at Ordinary Level examinations. The book should also prove particularly useful to students beginning Advanced Level General Certificate of Education and Ordinary National Certificate Science Courses.

In compiling the book, an attempt has been made to present relevant factual organic chemistry in a concise and logical fashion and to relate it to important natural and synthetic substances. Parts I and II of the book contain the factual chemistry which students will find essential in order to appreciate fully the role and scope of organic chemistry. Part III is concerned with the wider aspects of the subject and is intended to be of interest to elementary and advanced students alike.

At the end of each chapter in Parts I and II, experiments are described which are intended to assist the student to remember and understand the factual content of the book. It is recommended that the student performs these experiments and attempts to explain what is observed by reference to the text.

With regard to chemical nomenclature, it is advisable that at this level students should learn the names of common substances by example. At the same time, chemistry students should be aware that organic compounds can be named in a systematic way and for this reason both common and modern (I.U.P.A.C.) names are given with some word of explanation where appropriate.

A selection of test questions considered suitable for students preparing for the examinations mentioned above is included at the end of each chapter in Parts I and II.

D. E. F. ARMSTEAD

CONTENTS

PART III

INTRODUCTION

CHEMISTRY is a branch of science which is concerned with discovering the composition of all materials. The subject has been practised and studied for thousands of years but only with serious intent for about the last two hundred years. Prior to this there was very little scientific method as we understand it, the only chemical theories being those of alchemy and phlogiston. Alchemy was founded on the belief that common substances such as lead could be converted, by chemical means, into gold. The phlogiston theory maintained that all combustible materials contained phlogiston, a fire-producing agent, which they gave up when burned. These theories inspired men to investigate the properties of numerous substances, but because they were ill-conceived and never proven by experiment, no useful progress was made towards a better understanding of chemistry. It was not until these theories were finally discredited by men such as Lavoisier (in 1775) and Dalton (in 1808) that progress became possible and chemistry began to develop into the science studied today.

At a time when it seemed that all the old theories and superstitions had been cast aside and chemists were endeavouring to adopt a more rational approach to their work, another dispute arose as to whether organic substances could be synthesized by man. Many people at this time (the 1820s) believed that compounds of carbon, other than simple substances such as carbon dioxide, carbon monoxide, calcium carbonate and the carbides, could only be formed by the agency of a mysterious something called *vital force* which was only present in living things. The first synthesis of a truly organic substance was achieved by Wöhler in 1828. He showed that urea [$CO(NH_2)_2$], a substance found in urine, could be made from ammonium cyanate, which is an inorganic chemical. This discovery by Wöhler did much to dispel the idea that there was some important chemical difference between inorganic and organic substances; however, there were still many scientists and religious people who were unconvinced. It was not until about 1850, when other organic substances such as acetic acid and ethyl alcohol were synthesized, that the vital-force theory was disregarded. Today we realize that carbon is just one of 104 known elements and that general chemical principles apply to all these elements, including carbon. The only exceptional difference between carbon and other elements is the property of carbon to form an infinite variety of compounds as a result of the ease with which carbon atoms can link with one another to form chain and ring structures. As a result of this, the chemistry of carbon compounds is very extensive and requires special treatment if it is to be studied successfully.

This book serves simply as an introduction to organic chemistry, indicating the importance of the subject to our everyday lives and showing how organic compounds are classified according to their properties and uses.

PART I

1

HYDROCARBON COMPOUNDS

HYDROCARBON compounds contain hydrogen and carbon only.

THE ALKANES (Paraffins)

The simplest hydrocarbon compounds are the alkanes. These form a series in which methane is the first member. Methane has the formula CH_4. The atoms are covalently bonded to one another and the molecule has a tetrahedral structure.

At this stage, it should be noted that the bonding in all organic compounds is predominantly covalent. This term means that pairs of electrons are shared between atoms which are attached to one another in organic molecules. *Figure 1.1* (*a*) is a two-dimensional representation of the methane molecule and illustrates how electron pairs are shared between carbon and hydrogen atoms. *Figure 1.1* (*b*) is a three-dimensional representation of the methane molecule illustrating the directional nature of the bonds and the true shape of the molecule.

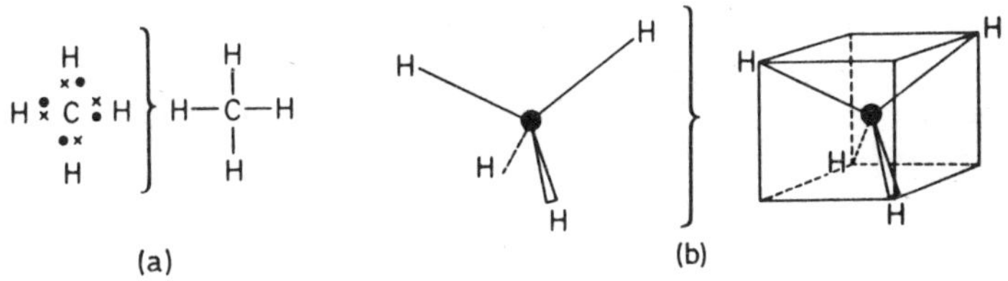

(a) (b)

Figure 1.1

The second member of the group is ethane, C_2H_6, the third member propane, C_3H_8, and the fourth member is butane, C_4H_{10}.

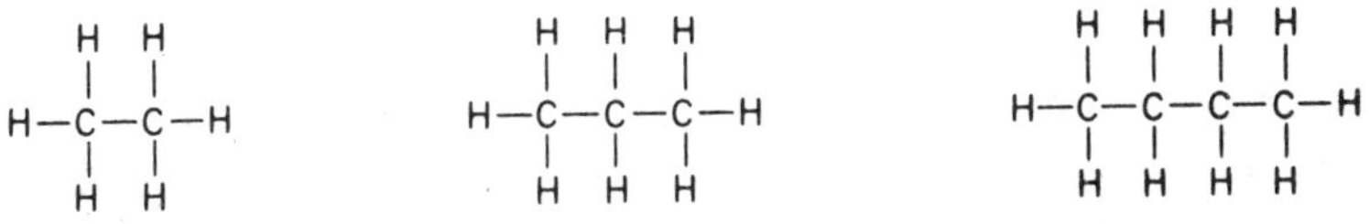

Ethane Propane Butane

These are structural formulae. They illustrate the arrangement of atoms relative to one another in the molecule. In propane, butane and the higher alkanes carbon maintains a tetrahedral structure and as a result molecules containing three or more carbon atoms have a zig-zag rather than a perfectly linear structure.

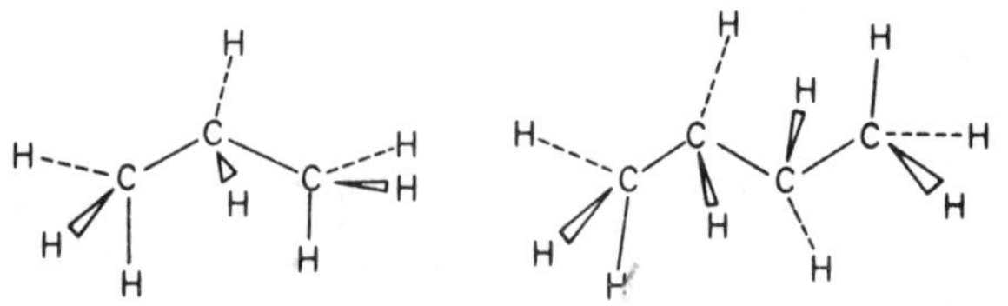

Propane

Butane

----- = behind plane of paper
—— = in plane of paper
▭▶ = in front of plane of paper

It will be noticed that the structural difference between methane and ethane is that the latter has one $-CH_2-$ more than the former, similarly propane has a $-CH_2-$ more than ethane and butane one more than propane. The alkane series of hydrocarbons is built up by successively introducing the $-CH_2-$ unit. The series is an example of a 'homologous series' (Greek: *homos*, same; *logos*, form) so called because the members show very similar physical and chemical properties.

Table 1.1 lists a few of the alkanes with their melting points and their boiling points. The first four members are named unsystematically but beginning with pentane their names are systematic and indicate the number of carbon atoms in their molecules and also that they belong to the alkane series. A small *n* in front of the name means that the carbon atoms in the hydrocarbon molecule are linked one to another in a 'straight chain'. In the case of butane and the higher-molecular-weight alkanes there are alternative ways in which the carbon atoms may be linked. Pentane, for example, can exist in three different structural forms. These different arrangements are called *isomers*.

The molecules of 2-methylbutane and 2:2-dimethylpropane are 'branched chain' molecules.

n–Pentane 2–Methylbutane (isoPentane) 2:2–Dimethylpropane (neoPentane)

Table 1.1

Alkane	*Formula*	m.p. °C	b.p. °C
Methane	CH_4	−184	−161
Ethane	C_2H_6	−172	− 88
Propane	C_3H_8	−190	− 45
n-Butane	C_4H_{10}	−135	1
n-Pentane	C_5H_{12}	−131	36
n-Hexane	C_6H_{14}	− 94	69

Preparation of Alkanes

General Methods

1. Alkanes may be prepared from alkyl halides (Chapter 8).

(a) By reduction with a zinc–copper couple in alcohol

e.g.
$$\underset{\text{ethyl iodide}}{C_2H_5I} + \underset{\text{nascent hydrogen}}{2[H]} \rightarrow \underset{\text{ethane}}{C_2H_6} + HI$$

The Zn–Cu couple is made by covering zinc turnings with copper sulphate solution. After a few minutes the copper sulphate solution is decanted leaving zinc coated with copper. The couple reacts with alcohol (+5% water) producing hydrogen and an organo-zinc compound.

(b) By reaction with sodium metal suspended in dry ether

e.g.
$$\underset{\text{ethyl iodide}}{C_2H_5\boxed{I + 2Na + I}C_2H_5} \rightarrow \underset{\text{butane}}{C_4H_{10}} + 2NaI$$

This type of reaction is known as a WURTZ synthesis. It is little used in preparative organic chemistry today and has the disadvantage of only being suitable for the preparation of alkanes containing an even number of carbon atoms.

2. Decarboxylation: Alkanes may be obtained from the sodium salts of monocarboxylic acids (p. 45) by heating with soda lime (i.e. a mixture of sodium hydroxide and calcium hydroxide)

e.g. $$\underset{\text{sodium propionate}}{C_2H_5COONa} + NaOH \rightarrow \underset{\text{ethane}}{C_2H_6} + Na_2CO_3$$

3. Kolbe's method: This is an electrolytic method. A concentrated aqueous solution of the sodium salt of a monocarboxylic acid is electrolysed between platinum electrodes
e.g.

$$2H_2O + \underset{\text{sodium acetate}}{2CH_3COONa} \rightarrow \underbrace{C_2H_6\uparrow + 2CO_2\uparrow}_{\text{at the ANODE}} + \underbrace{2Na^+ + 2OH^- + H_2\uparrow}_{\text{at the CATHODE}}$$

Preparation of Methane—In the laboratory, methane is conveniently obtained by adding water to aluminium carbide.

$$Al_4C_3 + 12H_2O \rightarrow 3CH_4 + 4Al(OH)_3$$

Properties

The alkanes are said to be 'saturated' indicating that the atoms comprising their molecules are linked by *single* covalent bonds. As a result of this, the alkanes are very stable and chemically rather inert, but under suitable conditions they will react with halogens and oxygen.

Methane combines with chlorine in the presence of sunlight to yield a variety of products:

$$CH_4 + Cl_2 \rightarrow \underset{\text{methyl chloride}}{CH_3Cl} + HCl$$

$$CH_3Cl + Cl_2 \rightarrow \underset{\text{methylene dichloride}}{CH_2Cl_2} + HCl$$

$$CH_2Cl_2 + Cl_2 \rightarrow \underset{\text{chloroform}}{CHCl_3} + HCl$$

$$CHCl_3 + Cl_2 \rightarrow \underset{\text{carbon tetrachloride}}{CCl_4} + HCl$$

With oxygen the alkanes react (burn) to form carbon dioxide and water:

$$\underset{\text{ethane}}{2C_2H_6} + 7O_2 \rightarrow 4CO_2 + 6H_2O + \text{Heat}$$

THE ALKENES (Olefines) AND ALKYNES (Acetylenes)

Some hydrocarbons contain carbon atoms held together by double and triple bonds. These compounds are said to be 'unsaturated' because unlike the alkanes some of the carbon atoms in the molecules do not take on their full quota of hydrogen.

Ethene (ethylene) contains a double bond and is the simplest unsaturated hydrocarbon.

C_2H_4 molecular formula; structural formula:

$$\begin{matrix} H & & & & H \\ & \diagdown & & \diagup & \\ & & C{=}C & & \\ & \diagup & & \diagdown & \\ H & & & & H \end{matrix}$$

The simplest unsaturated hydrocarbon containing a triple bond is ethyne (acetylene).

C_2H_2 molecular formula; $H{-}C{\equiv}C{-}H$ structural formula

Hydrocarbons containing carbon–carbon double bonds are collectively known as *alkenes* (old name, olefines) and hydrocarbons containing carbon–carbon triple bonds are known as *alkynes* (old name, acetylenes).

Although, in the structural formula above for ethene, two parallel straight lines are used to represent a carbon–carbon double bond, the bond is not equivalent to two ordinary single covalent bonds. Similarly the carbon–carbon triple bond is not equivalent to three single covalent bonds. From the values of bond energy and bond length given below it may be observed that the strength of a carbon–carbon double bond is less than twice the strength of a carbon–carbon single bond. In addition, carbon atoms joined by a double bond are closer together than when joined by a single bond. In the case of a triple bond the carbon atoms are even closer to one another and the bond is proportionately weaker.

$$>C{-}C<$$

(i) distance between carbon atoms = 1·54 Å*

(ii) energy, in kilocalories (1 kilocalorie = 1000 calories), required to break a carbon–carbon single bond = 80 per mole.

* Å = 1 ångström unit = 10^{-8} cm.

$$>C=C<$$

(i) distance between carbon atoms = 1·33 Å
(ii) energy required to break a carbon–carbon double bond = 145 kcal/mole.

$$—C\equiv C—$$

(i) distance between carbon atoms = 1·20 Å
(ii) energy required to break a carbon–carbon triple bond = 190·5 kcal/mole.

Preparation of Alkenes

General Methods

1. Dehydration of Alcohols

Excess concentrated sulphuric acid is heated with an alcohol

e.g. $$\underset{\text{ethanol}}{C_2H_5OH} + H_2SO_4 \xrightarrow{170°C} \underset{\text{ethyl hydrogen sulphate}}{C_2H_5HSO_4} + H_2O$$

$$C_2H_5HSO_4 \xrightarrow{170°C} \underset{\text{ethene}}{C_2H_4} + H_2SO_4$$

In this preparation ethene (ethylene) is always contaminated with sulphur dioxide from the decomposition of sulphuric acid and with carbon dioxide from the degradation of ethanol. Both of these impurities may be removed by passing the gas through sodium hydroxide solution.

2. Kolbe's Method

If a concentrated solution of the sodium salt of a dicarboxylic acid (i.e. one containing two carboxyl groups per molecule) is electrolysed between platinum electrodes, an alkene is produced at the anode

		ANODE	CATHODE
e.g. CH_2COONa		CH_2COO^-	$2Na^+$
$\vert$	$\xrightarrow{\text{electrolysis}}$	$\vert$	
CH_2COONa sodium succinate		CH_2COO^-	
		$\downarrow$	
		$CH_2\uparrow + 2\overrightarrow{CO_2}$	
		$\Vert$	
		CH_2 ethene	

Note: This method only applies when the two carbon atoms attached to the carboxylic acid groups are adjacent to one another (as for example in sodium succinate).

Preparation of Alkynes

General Methods

1. Dehydrohalogenation of a vic-dihalide (vic: abbreviation for vicinal, denoting that the halogen atoms are on adjacent carbon atoms)

e.g.

$$\underset{\text{1:2-dibromoethane}}{BrCH_2CH_2Br} + \underset{\substack{\text{dissolved in}\\\text{alcohol}}}{KOH} \longrightarrow \begin{array}{c}CH_2\\ \|\\ CHBr\end{array} + KBr + H_2O$$

$$\underset{\text{vinyl bromide}}{\begin{array}{c}CH_2\\ \|\\ CHBr\end{array}} + KOH(alc) \rightarrow \underset{\text{ethyne (acetylene)}}{\begin{array}{c}CH\\ |||\\ CH\end{array}} + KBr + H_2O$$

2. Kolbe's method

An alkyne can be made by electrolysing a solution of the sodium salt of an unsaturated dicarboxylic acid

e.g.

$$\underset{\text{sodium fumarate}}{\begin{array}{c}CHCOONa\\ \|\\ CHCOONa\end{array}} \rightarrow \overset{\text{ANODE}}{\begin{array}{c}CHCOO^-\\ \|\\ CHCOO^-\end{array}} \quad \overset{\text{CATHODE}}{2Na^+}$$

$$\downarrow$$

$$\underset{\text{ethyne (acetylene)}}{\begin{array}{c}CH\uparrow\\ |||\\ CH\end{array}} + 2\overrightarrow{CO}_2$$

Preparation of Ethyne (Acetylene)

Ethyne may conveniently be prepared in the laboratory by adding water to calcium carbide.

$$CaC_2 + 2H_2O \rightarrow \underset{\text{ethyne}}{C_2H_2\uparrow} + Ca(OH)_2$$

Properties of Alkenes and Alkynes

Alkenes and alkynes are very reactive. They undergo numerous reactions in which the double and triple bonds between adjacent carbon atoms are partially broken. The most common reactions of these compounds are *addition reactions*.

1. Unsaturated compounds react with halogens to form *addition compounds*

e.g. $CH_2{=}CH_2 + Br_2(\text{liquid}) \rightarrow CH_2BrCH_2Br$
ethene — 1:2-dibromoethane

$HC{\equiv}CH + 2Br_2(\text{liquid}) \rightarrow CHBr_2CHBr_2$
ethyne — 1:1:2:2-tetrabromoethane

The double and triple bonds between adjacent carbon atoms are broken and bromine atoms add on to each carbon atom to satisfy the latter's quadrivalency.

2. Acidified potassium permanganate solution is decolorized by unsaturated compounds. This is due to *hydroxylation*, i.e. addition of hydroxyl groups across the double bond

e.g. $CH_2{=}CH_2 + H_2O + \hat{O}(\text{from } KMnO_4) \rightarrow CH_2(OH)CH_2(OH)$
ethene — 1:2-dihydroxyethane (glycol)

$CH{\equiv}CH + 2\hat{O}(\text{from } KMnO_4) \rightarrow CHOCHO$
ethyne — ethanedial (glyoxal)

$CHOCHO + \hat{O} + H_2O \rightarrow 2HCOOH$
methanoic acid (formic acid)

3. Alkenes and alkynes will combine with halogen acids to form addition compounds

e.g. $CH_2CH_2 + HI \rightarrow CH_3CH_2I$
ethene — ethyl iodide

$CH{\equiv}CH + HI \rightarrow CH_2{=}CHI$
ethyne — vinyl iodide

$CH_2{=}CHI + HI \rightarrow CH_3CHI_2$
1:1-di-iodoethane

4. Double and triple carbon–carbon bonds add on hydrogen. In the case of ethene and ethyne (ethylene and acetylene) the gases are mixed with hydrogen and the mixture is passed over heated finely-divided nickel

e.g. $CH_2{=}CH_2 + H_2 \rightarrow CH_3CH_3$
ethene — ethane

$CH{\equiv}CH + H_2 \rightarrow CH_2{=}CH_2$
ethyne — ethene

and $CH_2{=}CH_2 + H_2 \rightarrow CH_3CH_3$

The process involving the addition of hydrogen is called *hydrogenation*.

The alkenes and alkynes are further examples of homologous series of compounds.

Alkenes	*Alkynes*
$CH_2{=}CH_2$ ethene	$CH{\equiv}CH$ ethyne
$CH_2{=}CH{-}CH_3$ propene	$CH{\equiv}C{-}CH_3$ propyne
$CH_2{=}CH{-}CH_2CH_3$ butene (but-1-ene), etc.	$CH{\equiv}C{-}CH_2{-}CH_3$ butyne (but-1-yne), etc.

BENZENE HYDROCARBONS

The hydrocarbons so far considered are examples of *aliphatic* compounds. Their molecules are linear, that is to say, the carbon atoms are joined in straight or branched chains

e.g. $CH_3CH_2CH_2CH_2CH_3$ straight chain
n-pentane

$$\begin{array}{c} \quad\; CH_3 \\ \quad\; | \\ CH_3CH_2CH\ CH_3 \end{array}$$ branched chain
2-methylbutane

Besides aliphatic organic compounds there are others known as *aromatic* compounds. The most common of these contain groups of six carbon atoms arranged in rings. Benzene, for example, is a simple aromatic compound and has the structural formula:

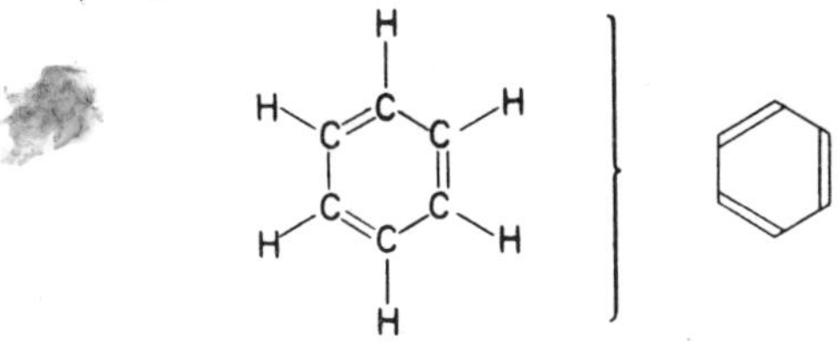

Structural formula Symbol

Naphthalene is a more complex aromatic compound and consists of two *fused* benzene rings:

Structural formula Symbol

Compounds containing benzene ring systems are unsaturated; the carbon to hydrogen ratio is high.

Properties of Benzene

Benzene has the molecular formula, C_6H_6. The structural formula of benzene was given above as:

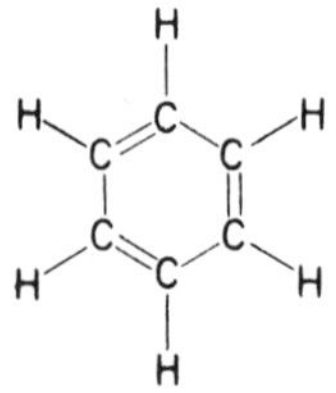

Although this representation satisfies the valency requirements of carbon and hydrogen, it does not fully account for the properties of benzene. By comparing the structure of ethene (p. 7) with that shown above for benzene we might expect the two compounds to have many chemical properties in common. This is not so, the benzene molecule is chemically more stable than the ethene molecule. This is shown in the following reactions of benzene.

1. Bromine will react with benzene to give a number of different products:

(a) In bright sunlight benzene forms an addition compound with bromine but *the reaction does not take place as readily as with ethene.*

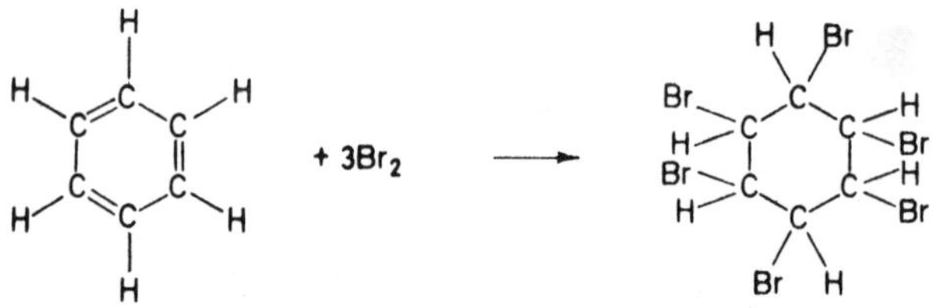

$+ 3Br_2 \longrightarrow$

Benzene hexabromide

(b) In the presence of a catalyst such as iron filings, bromine takes the place of hydrogen atoms in the ring. Reactions of this type are called *substitution* reactions; bromine being substituted for hydrogen. In this particular case the product formed will depend on the proportion of bromine and benzene used.

For the purpose of naming disubstitution products of benzene the carbon atoms in the ring are numbered. Consider the dibromobenzenes below. Compound I is named 1:2-dibromobenzene and

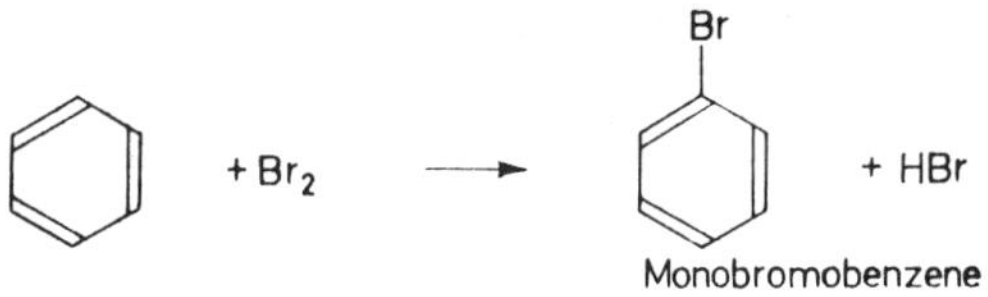

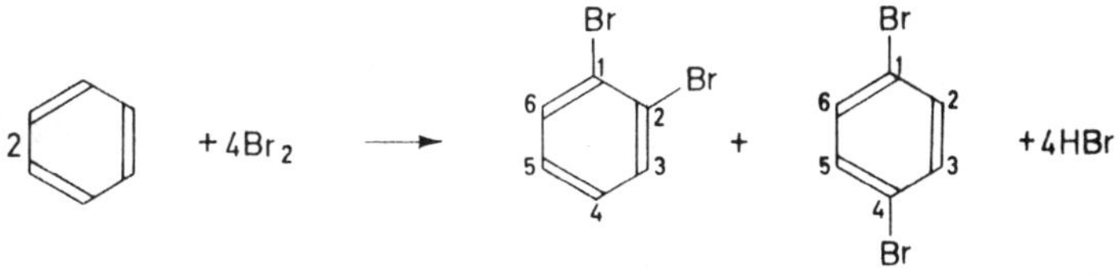

$$C_6H_6 + Br_2 \longrightarrow C_6H_5Br + HBr$$

Monobromobenzene

$$2C_6H_6 + 4Br_2 \longrightarrow \text{I} + \text{II} + 4HBr$$

I II

compound II 1:4-dibromobenzene. An alternative method of naming benzene homologues containing more than one substituent is by making use of the prefixes, *ortho(o)*, *meta(m)*, and *para(p)*. (This method is still in common use.) These prefixes refer to the following positions in the ring: *ortho-*, 2 and 6; *meta-*, 3 and 5; and *para*: 4. For example, 1:2-dibromobenzene becomes *o*-dibromobenzene and 1:4-dibromobenzene becomes *p*-dibromobenzene. *meta*-Dibromobenzene corresponds to 1:3-dibromobenzene.

2. Potassium permanganate does not react with benzene unless the oxidizing agent is present in excess and the mixture is boiled. Under these conditions benzene is eventually converted to carbon dioxide and water.

$$\underset{\text{(from KMnO}_4\text{)}}{C_6H_6 + 15\hat{O}} \xrightarrow{\text{HEAT}} 6CO_2 + 3H_2O$$

3. Benzene will not react with halogen acids (unlike ethene).

4. Benzene will react with hydrogen in the presence of a nickel catalyst at about 200°C to form *cyclo*hexane. The reaction takes place, proportionately, less readily than with ethene.

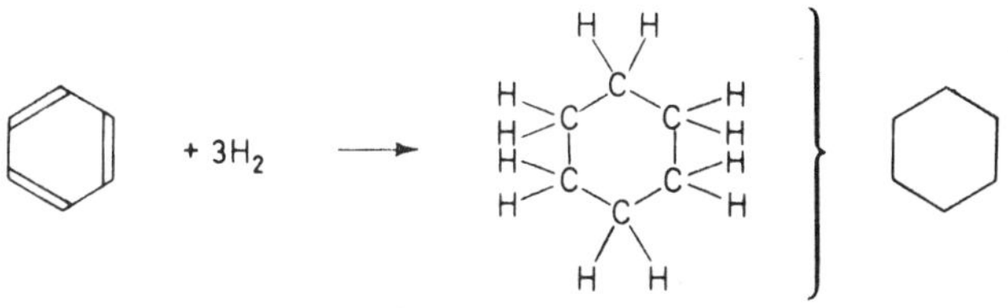

$$C_6H_6 + 3H_2 \longrightarrow C_6H_{12}$$

Cyclohexane Symbol

To explain the peculiar chemical nature of benzene a number of structures have been proposed. The most useful of these describes the benzene molecule as having the basic skeleton:

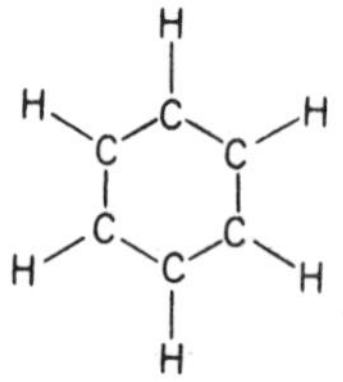

in which the atoms are joined by 'ordinary' covalent bonds referred to as 'sigma' (σ) bonds. This accounts for three of the four valency electrons belonging to each carbon atom. The remaining electrons are said to be 'delocalized' over the ring, forming electron clouds above and below the ring in the manner illustrated below.

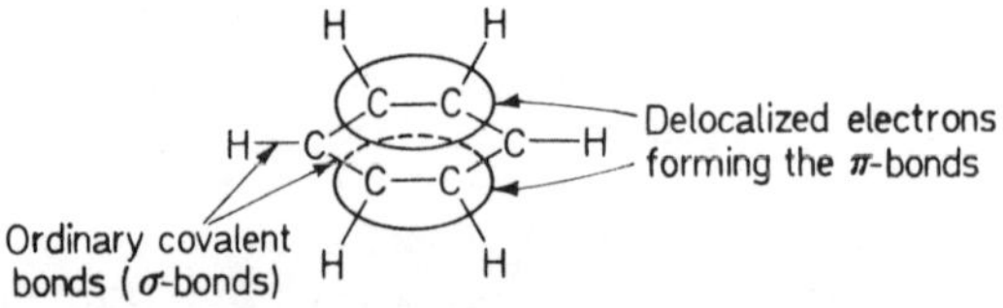

Delocalized electrons forming the π-bonds

Ordinary covalent bonds (σ-bonds)

The delocalized electrons, of which there are six, are referred to as 'pi' (π)-electrons and the bonds formed by a combination of these electrons are called π-bonds. The benzene structure is very stable and, as a result, does not readily undergo reactions which are generally typical of aliphatic unsaturated systems.

For the purpose of writing equations benzene is usually represented as shown on p. 11. This is simply a convenience since the structure shown above, although a more accurate representation, is rather cumbersome.

EXPERIMENTS

Preparation of Ethane

(A Würtz synthesis using copper in place of sodium.)

Put about 4 c.c. of methyl iodide into a hard glass test-tube and push sufficient asbestos wool into the tube to soak up the iodide. Add copper clippings to half fill the test-tube and then connect up as shown in *Figure 1.2*.

Heat the test-tube in the region of the copper clippings and collect the gas evolved.

Using a lighted taper ignite a tube of the gas and notice that it

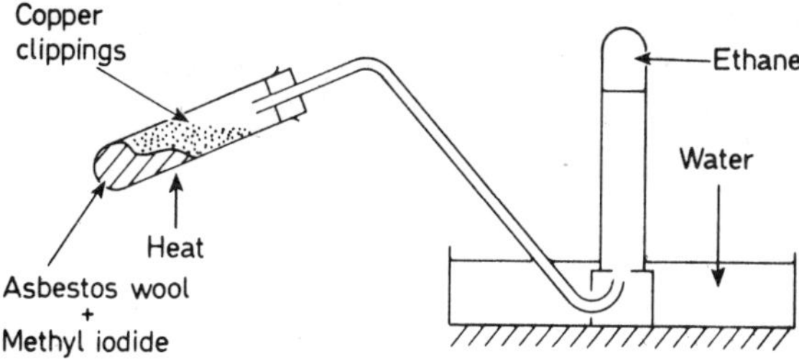

Figure 1.2

burns with a bluish flame. Write the equation for the combustion of ethane.

Preparation of Ethene (ethylene)

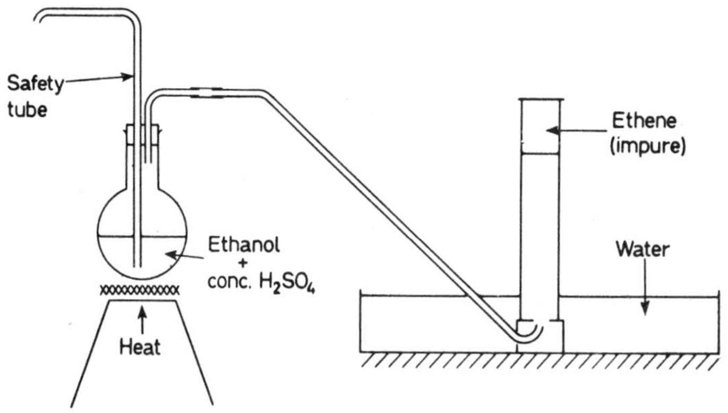

Figure 1.3

Introduce 10 ml. of ethanol into a 100 ml. flask and slowly add 20 ml. of concentrated sulphuric acid. It is necessary to cool the mixture, as concentrated acid is added, by running cold water over the outside of the flask. Connect up the apparatus as shown in *Figure 1.3* and heat the contents of the flask. (*Caution:* Heat the flask gently and when the reaction mixture turns dark brown remove the heat source. If necessary re-heat gently to maintain the evolution of ethene.) Collect the gas over water.

Shake gas jars of ethene with (i) a very dilute solution of alkaline potassium permanganate (made by adding an equal volume of dilute sodium carbonate solution to a few millilitres of dilute potassium permanganate solution), and (ii) bromine water. Record what occurs and write equations for the reactions. *Note:* The gas obtained in the manner described above will be contaminated with sulphur dioxide and carbon dioxide. Suggest methods of removing these impurities.

Preparation of Ethyne (Acetylene)

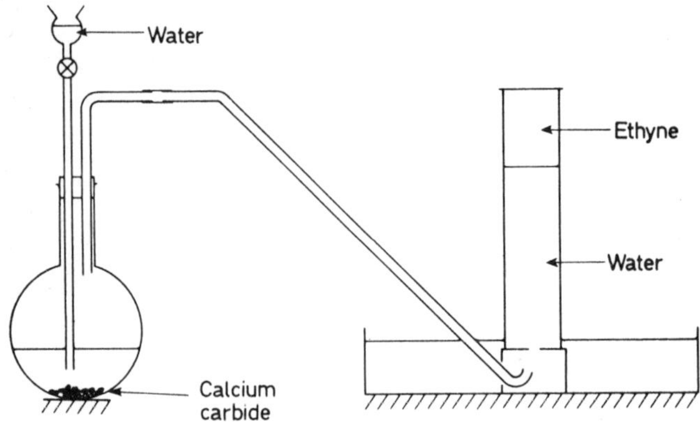

Figure 1.4

Using the apparatus shown in *Figure 1.4*, add water, dropwise, to calcium carbide and collect the gas over water. Ignite a test-tube of the gas and notice the colour of the flame. Why does ethyne burn with a different coloured flame to that of ethene?

QUESTIONS

1. What is a hydrocarbon compound? By writing structural formulae give examples of saturated and unsaturated hydrocarbon compounds.
2. Explain the meaning of the term homologous series.
3. Describe *TWO* general methods of preparing alkanes. Write equations for the reactions you mention.
4. How may methane be prepared in the laboratory? Draw the apparatus you would use to prepare the pure, dry gas.
5. What are the possible products from the reaction between chlorine and ethane? Write equations for the reactions and name the products.
6. Calculate the volume (measured at N.T.P.) of oxygen required to oxidize completely 2 gram molecules of ethane. (The molecular weight of any gas at N.T.P. occupies 22·4 litres.)

7. Describe, in detail, the laboratory preparation of ethene (ethylene). How does it react with bromine and alkaline potassium permanganate solution?
8. Describe the laboratory preparation of a pure specimen of ethyne (acetylene).
9. Compare the structure and simple chemical properties of benzene and ethene.
10. Write the structural formulae of the following named substances: (a) 2-methylbutane, (b) 2-methyl-4-chlorohexane, (c) 3-*iso*-propylpentane (i.e. 3-ethyl-2-methylpentane).

2

NATURAL GAS AND PETROLEUM

COMPOSITION AND OCCURRENCE

NATURAL gas is a mixture of methane (60–80%), ethane (5–9%), propane (3–18%) and higher-molecular-weight hydrocarbons (2–14%). Petroleum is an extremely complex liquid mixture of hydrocarbons, it is estimated for example, that petroleum boiling up to 200°C contains at least 500 organic compounds.

Natural gas and petroleum are obtained from wells drilled into porous rocks containing the gas and oil. These wells may be up to two miles deep, and usually gas and oil occur together although in some parts of the world they are found separately. The gas and oil are thought to be derived from decayed animal and vegetable matter.

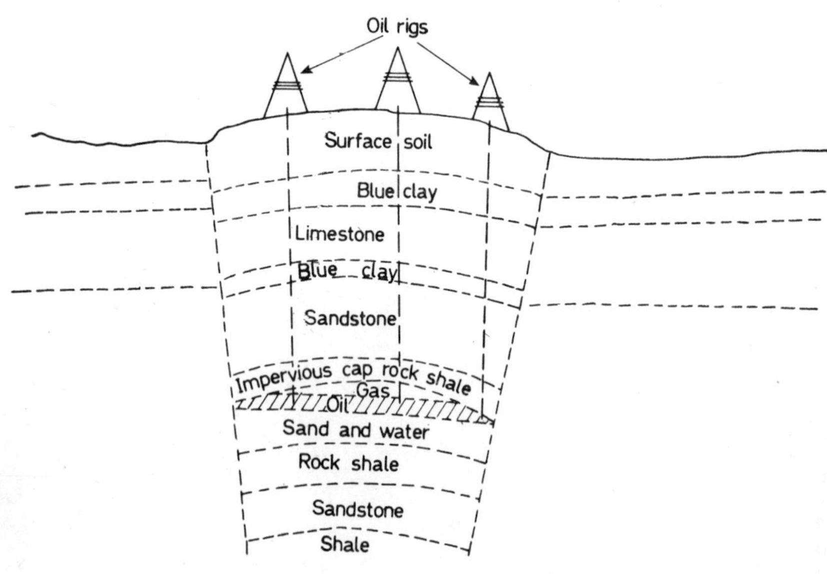

Figure 2.1. *Oil Field*

REFINING OF PETROLEUM

The refining of petroleum is essentially a two-stage process entailing cleaning and separation of the components by distillation.

Cleaning

The oil is filtered to remove solid matter and then desalted. Desalting involves the removal of sodium chloride and other water-soluble salts by washing with water at about 80°C under pressure.

Distillation

After inorganic substances have been removed from the oil, it is distilled to effect a partial separation of the organic constituents. Distillation entails boiling the liquid mixture and condensing and collecting the vapour. The principle is that the constituents of the mixture having the lowest boiling points vaporize and condense first.

The method of distillation used in this case is referred to as 'fractional distillation'. This differs from simple distillation in that, placed between the distillation flask and the condenser is a 'fractionating column'. This device facilitates separation of the mixture into a number of fractions, each comprising relatively few components.

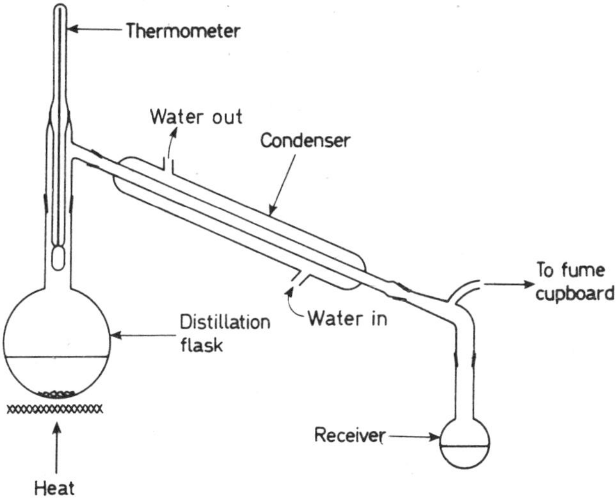

Figure 2.2. *Apparatus for simple distillation*

The commercial, large-scale, fractionating column is designed in such a way as to enable various fractions to be tapped off directly

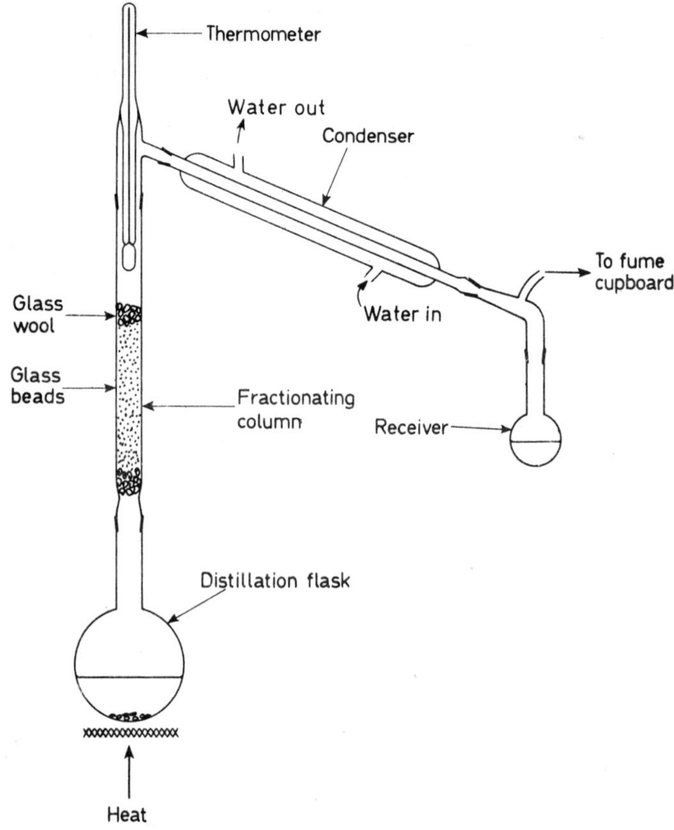

Figure 2.3. *Apparatus for fractional distillation*

from the column (*Figure 2.4*). Preliminary distillation of petroleum usually separates the oil into six main fractions.

Table 2.1

Fraction	*Boiling range,* °C
1. Gas	below 50
2. Petrol	40–175
3. Kerosine	150–240
4. Light oil	220–250
5. Heavy oil	250–350
6. Residue	above 350

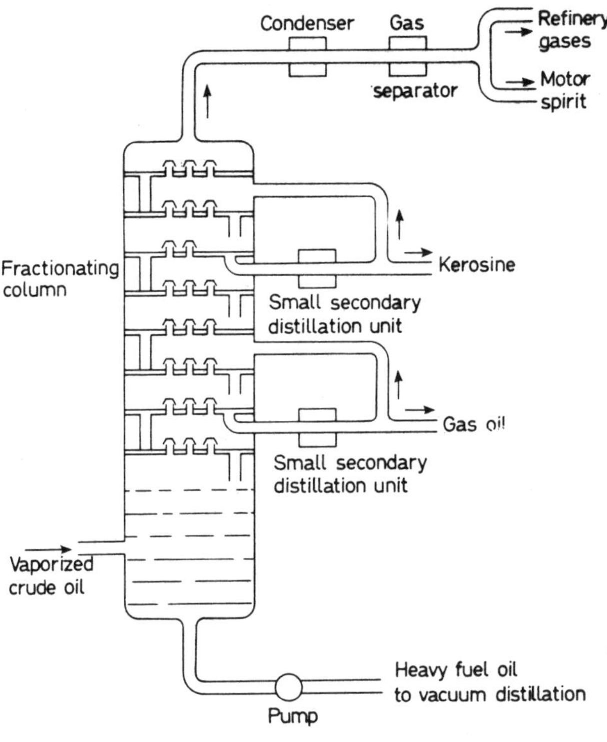

Figure 2.4

Fraction 1

The gas fraction consists of simple hydrocarbons and is mostly used as a fuel.

Fraction 2

The petrol fraction consists of higher-molecular-weight hydrocarbons (hydrocarbons containing 4 to 10 carbon atoms per molecule) and is used as a motor fuel.

The petrol fraction from the primary distillation is not used directly. It must first be treated in a process called *reforming* before it is suitable for use in an automobile engine. The primary distillation product consists of 'straight chain' hydrocarbons and by reforming these are converted to 'branched chain' hydrocarbons.

E.g. $CH_3CH_2CH_2CH_2CH_3 \rightarrow CH_3CH_2{\cdot}CH(CH_3){\cdot}CH_3$

n-pentane → 2-methylbutane

Reforming is achieved by heating crude petrol to about 500°C under pressure in the presence of a molybdenum-on-alumina catalyst. The lower-molecular-weight straight chain hydrocarbons are converted into branched chain molecules as illustrated by the example above. Higher-molecular-weight hydrocarbons are 'cyclized' and dehydrogenated (hydrogen removed) to aromatic compounds (p. 11)

e.g. $CH_3CH_2CH_2CH_2CH_2CH_2CH_3 \longrightarrow$

n-heptane

(methylcyclohexane ring: CH_3–CH, H_2C, CH_2, H_2C, CH_2, CH_2) $\xrightarrow{-3H_2}$ (benzene ring with CH_3)

Toluene

In addition to reforming, the branched nature of the components of the petrol fraction may be increased by *cracking*, *alkylation*, *isomerization* and *polymerization*.

Cracking—This is the thermal decomposition of hydrocarbon compounds

e.g. $CH_3(CH_2)_8CH_3 \xrightarrow[500°C]{SiO_2/Al_2O_3} CH_3CH{=}CH_2 + C_2H_4 + C_4H_8 + CH_4$

n-decane; propene; ethene; butene; methane

Alkylation—This is a method for introducing alkyl groups into compounds such as alkenes

e.g. $CH_3CH{=}CH_2 + CH_3CH(CH_3)CH_3 \xrightarrow[\text{or } SiO_2/Al_2O_3]{AlCl_3} CH_3CH(CH_3)CH_2CH(CH_3)CH_3$

propene; 2-methylpropane (*iso*butane); 2:4-dimethylpentane

Isomerization—This is a process in which a straight chain hydrocarbon is converted into a branched hydrocarbon e.g. $$2CH_3CH_2CH_2CH_2CH_2CH_3 \xrightarrow[AlCl_3]{300^\circ C} CH_3CH(CH_3)CH_2CH_2CH_3 + CH_3CH_2CH(CH_3)CH_2CH_3$$

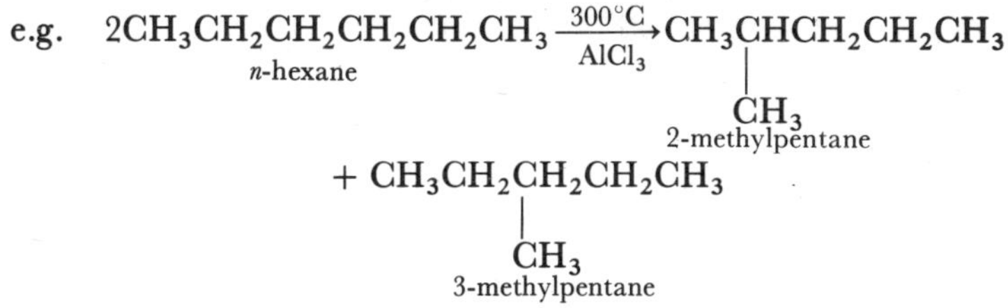

n-hexane → 2-methylpentane + 3-methylpentane

Polymerization—In this process small molecules are joined together to give hydrocarbons having higher octane numbers

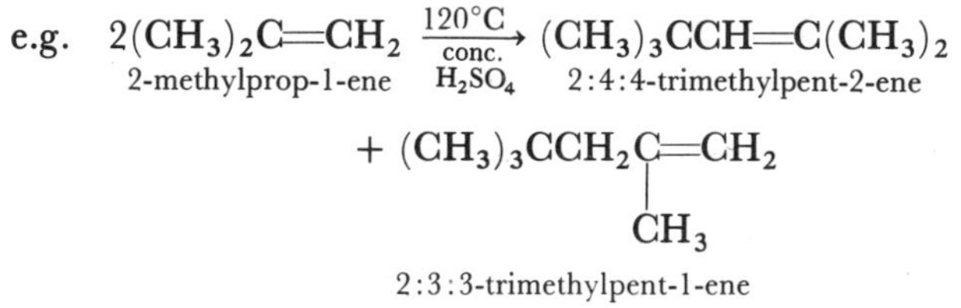

e.g. $$2(CH_3)_2C{=}CH_2 \xrightarrow[\text{conc. } H_2SO_4]{120^\circ C} (CH_3)_3CCH{=}C(CH_3)_2 + (CH_3)_3CCH_2C(CH_3){=}CH_2$$

2-methylprop-1-ene → 2:4:4-trimethylpent-2-ene + 2:3:3-trimethylpent-1-ene

The term *octane number* is used to denote the branched nature and degree of cyclization of petrol fractions used as motor fuels. It has been found that 2:2:4-trimethylpentane (incorrectly named *iso*-octane) burns in an internal combustion engine without causing a metallic rattle referred to as *knocking*. On the other hand, *n*-heptane because of its straight-chain nature gives rise to knocking and is unsuitable as a petrol engine fuel. 2:2:4-Trimethyl pentane is arbitrarily given the number 100 and *n*-heptane the number 0. Petrol fractions are compared with 2:2:4-trimethylpentane and *n*-heptane mixtures and given a number (grade), on the arbitrary scale 100–0, referred to as the octane number. Generally, the higher

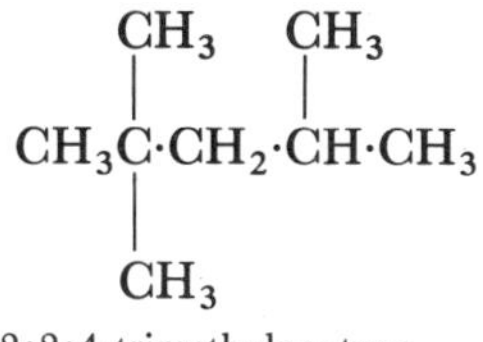

$$CH_3C(CH_3)_2{\cdot}CH_2{\cdot}CH(CH_3){\cdot}CH_3$$

2:2:4-trimethylpentane

$$CH(CH_3)_2{\cdot}CH_2CH_2CH_2CH_2CH_3$$

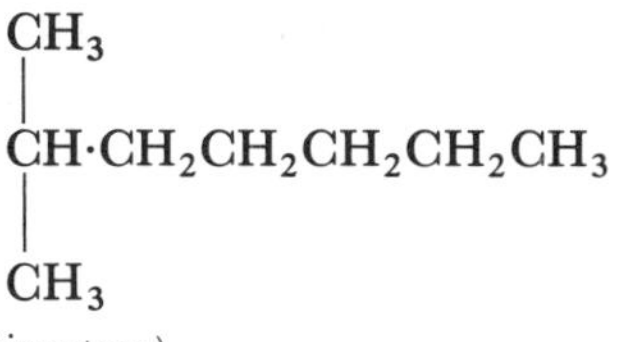

(*iso*octane)

$$CH_3CH_2CH_2CH_2CH_2CH_2CH_3$$

n-heptane

the octane number the greater the branched nature of the petrol.

Fraction 3

The kerosine fraction consists of hydrocarbons containing 10–15 carbon atoms per molecule. The oil is used as an aeroplane fuel and as a domestic fuel.

As a jet engine fuel, the oil must be of the correct viscosity in order to ensure efficient atomization (to produce a fine spray) at high altitudes where the air temperature may be below −80°C.

Fraction 4

This fraction is of still higher molecular weight, it consists of hydrocarbons containing 15–20 carbon atoms per molecule.

At the present time the demand for straight run kerosine and light oil does not exceed the supply, there are far more of these fractions than is required. Much of the excess is converted to lower-molecular-weight hydrocarbons by cracking. This involves heating kerosine and light oil to a high temperature (500°–600°C) under pressure. Under these conditions long-chain hydrocarbons are broken into smaller ones. The products of cracking depend on the conditions of operation, i.e. catalysts, temperature and pressure. Before about 1940 cracking consisted simply of heating the oil to a high temperature under pressure, but as a result of World War II and the increased demand for high-octane fuels a more efficient process was developed using catalysts (e.g. aluminium silicates).

Fraction 5

Heavy oil contains hydrocarbons having 20–25 carbon atoms per molecule. It is used as a fuel in diesel engines.

Fraction 6

The residue is distilled under reduced pressure to obtain lubricating oils, greases, waxes and bitumen.

The oils and greases are used as engine lubricants and the waxes are used to manufacture polishes, candles and waxed paper. Bitumen is used for road making and for insulation.

SOME IMPORTANT COMPOUNDS OBTAINED FROM PETROLEUM

Methane CH_4

Natural gas consists mostly of methane. Methane is used as a fuel (fed into the coal gas supply) and for making methanol (CH_3OH), ammonia and carbon black. Methanol is used as a solvent and in the manufacture of plastics. Carbon black is used for making black inks.

$$CH_4 + H_2O \text{ (steam)} \xrightarrow[900°C]{Ni} CO + 3H_2$$

$$CO + 2H_2 \xrightarrow[Zn/Cr \text{ oxides}]{400°C} CH_3OH$$

$$CH_4 \xrightarrow{1200°C} C + 2H_2$$

Ethene $CH_2{=}CH_2$

This gas may be obtained from petrol by heating with steam at about 750°C for a short time. Ethane, propene and other hydrocarbons are also formed.

Ethene is very important since a large number of chemicals are made from it, i.e. plastics, ethanol and ethene oxide.

$$C_2H_4 \xrightarrow[200°C,\ 200 \text{ atm}]{Catalyst} \underset{\text{polythene}}{(—CH_2—CH_2—)_n} \qquad n = 10{,}000–50{,}000$$

$$C_2H_4 + H_2SO_4 \rightarrow \underset{\text{ethyl hydrogen sulphate}}{C_2H_5HSO_4} \xrightarrow{H_2O} \underset{\text{ethanol}}{C_2H_5OH} + H_2SO_4$$

$$C_2H_4 + \tfrac{1}{2} O_2 \xrightarrow[250°C]{Ag} \underset{\text{ethene oxide}}{CH_2—CH_2 \text{ (bridged by O)}} \xrightarrow{H_2O} \underset{\text{1:2-dihydroxyethane}}{CH_2OH—CH_2OH}$$

Ethyne $CH{\equiv}CH$

This gas can be obtained from methane by heating to about 1500°C for a short time.

$$2CH_4 \rightarrow C_2H_2 + 3H_2$$

Note: In the past, ethyne was produced in the U.K. from calcium carbide.

$$CaC_2 + 2H_2O \rightarrow C_2H_2 + Ca(OH)_2$$

Ethyne is used to make tetrachloroethane, vinyl chloride, plastics and acetaldehyde (not a great deal of acetaldehyde is made from ethyne today).

$$CH{\equiv}CH + 2Cl_2 \xrightarrow[Catalyst]{Light} \underset{\text{tetrachloroethane}}{CHCl_2CHCl_2}$$

$$CH{\equiv}CH + HCl \xrightarrow[65°C,\ Hg^{2+}]{H_2O} \underset{\text{vinyl chloride}}{CH_2{=}CHCl}$$

$$CH{\equiv}CH \xrightarrow[FeSO_4 + HgSO_4]{\text{Dilute } H_2SO_4} \underset{\text{acetaldehyde}}{CH_3CHO}$$

Aromatic Compounds

Benzene, toluene and the xylenes are amongst the most important aromatic compounds derived from petroleum. The importance of these aromatic compounds may be judged from their variety of uses (*Table 2.2*).

CH_3 CH_3 CH_3 CH_3 CH_3 CH_3 CH_3

Benzene Toluene *ortho*-Xylene *meta*-Xylene *para*-Xylene

Table 2.2

Benzene	*Toluene*	*Xylenes*
Solvent	Solvent	Solvent
Fuel	Fuel	Fuel
For making:	For making:	For making:
*cyclo*Hexane	T.N.T.	Phthalic acid
Styrene and polystyrene	Benzyl alcohol	(for dyes and plastics)
Benzenehexachloride	Benzaldehyde	Terephthalic acid
Isododecyl	Benzoic acid	(for making terylene)
(for making detergents)	Saccharin	

EXPERIMENTS

Determination of Boiling Points and Melting Points

Boiling Points

The boiling point of a liquid available in large quantity is best determined using the simple distillation apparatus shown in *Figure 2.2*. The liquid is distilled and the constant temperature of distillation is taken as the boiling point of the liquid.

If only a small amount of the liquid is available, then the following method is recommended. The liquid (e.g. an alkane) is put into an ignition tube and the tube incorporated in the apparatus shown in *Figure 2.5*. The liquid paraffin is heated gently and when the boiling point is reached, bubbles of air issue in a *steady stream* from the open end of the capillary tube. The temperature at which this occurs is recorded as the boiling point of the liquid in the ignition tube.

Melting Points

In this case, the crystalline substance (e.g. naphthalene) is introduced

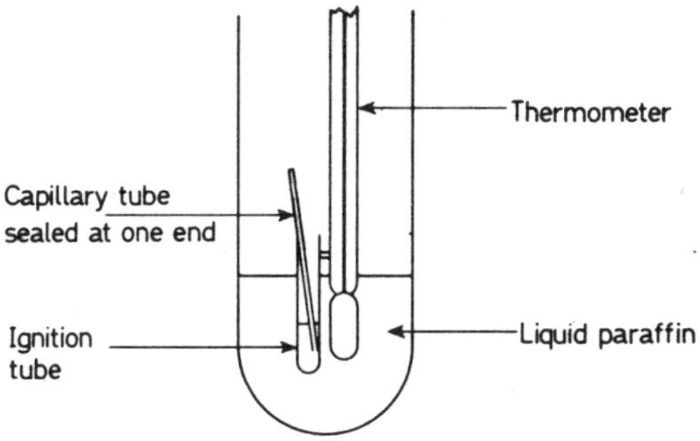

Thermometer
Capillary tube sealed at one end
Ignition tube
Liquid paraffin

Figure 2.5

into a capillary tube closed at one end and the tube is attached to the bulb of the thermometer by an elastic band. The bulb of the thermometer is then immersed in liquid paraffin contained in a boiling tube and the latter heated gently. The solid turns to liquid at the m.p.

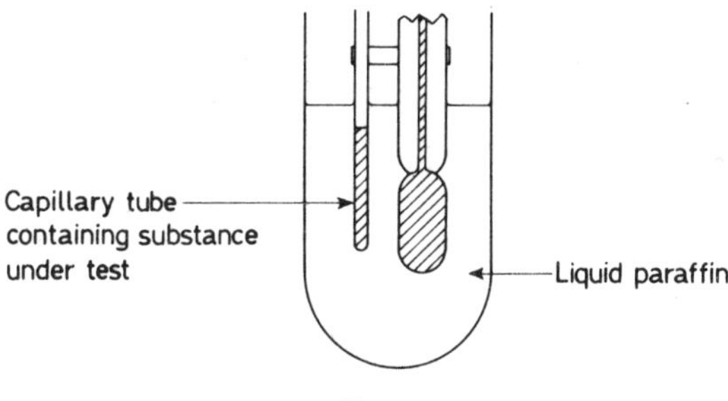

Figure 2.6

Cracking

Set up the apparatus shown in *Figure 2.7*. Heat the catalyst strongly and then allow the alkane to pass slowly into the combustion tube (heated sufficiently to keep the catalyst hot). Collect the gas and test it as follows:

(i) investigate the combustion properties of a test-tube full of the gas;

(ii) shake the gas with (a) bromine water, and (b) alkaline potassium permanganate solution.

Record and explain what happens.

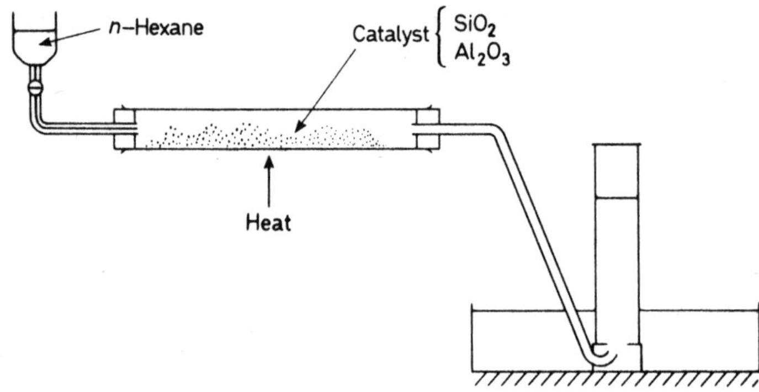

Figure 2.7

QUESTIONS

1. What is the approximate composition of natural gas and petroleum?
2. Explain how crude oil is refined.
3. Draw labelled diagrams of the apparatus used for simple and fractional distillation.
4. Why is it that petrol, obtained by fractional distillation of crude petroleum, is unsuitable for direct use in an automobile engine?
5. Tabulate the important chemicals obtained from petroleum and mention their uses.
6. Explain very briefly the meaning of the following terms: (a) reforming, (b) cracking, (c) polymerization, (d) isomerization, and (e) octane number.

3

COAL

INTRODUCTION

COAL is a composite mixture of substances containing carbon, hydrogen, oxygen, nitrogen, sulphur and small proportions of other elements. It is derived from decayed animal and vegetable matter which has been subjected to high pressures and temperatures resulting from movements of the earth's crust. The formation of carbonaceous material (such as coal) from plants and animals is a continuous process involving a very large number of chemical changes. We may summarize these changes as follows.

The fibrous part of plants consists mostly of cellulose which is a complex carbohydrate (see p. 84). Cellulose is made by plants from carbon dioxide and water in a process known as *photosynthesis*. Carbon dioxide passes into the plant leaves from the air and water is absorbed from the soil through the plant roots.

$$CO_2 + H_2O \xrightarrow[\text{sunlight}]{\text{catalyst}} \text{sugars} + O_2\uparrow$$
$$\downarrow$$
$$\text{cellulose}$$

The catalyst for this reaction is a substance called chlorophyll. It is this which gives plants their green colour. In addition to chlorophyll the plant also requires energy in the form of sunlight to convert carbon dioxide and water to cellulose. As indicated in the equation above, the conversion is not a simple one. It involves numerous chemical reactions and the formation of many simple substances such as sugars (e.g. glucose) before cellulose is formed. Oxygen is liberated in the process and replaces carbon dioxide from the air.

Plants eventually die and decay and if the conditions are right, their remains, over a period of many thousands of years, may form coal. If the coal is now burned in air (i.e. oxidized) carbon dioxide, water and a great deal of heat are produced. These are the very materials which plants use to make cellulose and which is ultimately transformed into coal. The heat produced when coal is combusted is another form of the radiant energy derived by plants from the sun.

The whole sequence of changes carbon undergoes in natural processes is known as the *carbon cycle* and may be represented diagrammatically as shown in *Figure 3.1*.

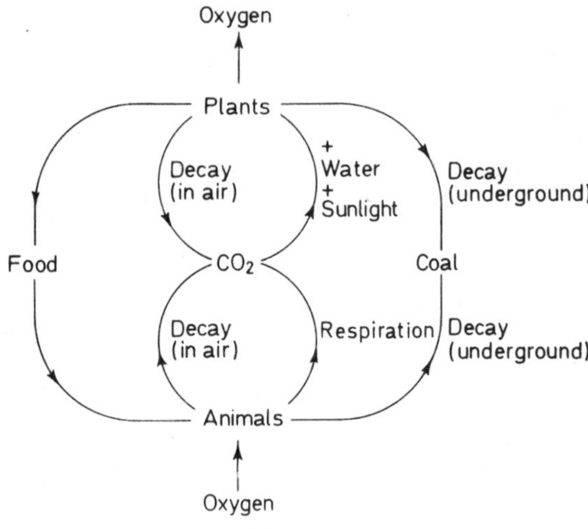

Figure 3.1

The composition of coal varies depending on how it has been formed and its age. The oldest coals are very hard and contain a high proportion of carbon whereas the more recent coals are relatively soft and contain, besides carbon, a high proportion of water and volatile substances. The approximate composition of three commonly-occurring coals are shown in *Table 3.1*.

Table 3.1

Type of coal	*Ash,* %	*Fixed carbon,* %	*Volatile matter,* %	*Moisture* %	*Calorific value/g*
Lignite	8	43	43	6	7000
Bituminous	6	61	29	4	8000
Anthracite	5	87	6	2	9000

Coal is important as a source of fuel gas, fertilizers, organic chemicals, sulphuric acid and coke, but since 1945 its importance as a source of organic chemicals has diminished. In 1945 about 95% of

Britain's organic chemicals were obtained from coal but today petroleum is the main source and coal only accounts for about 30% of the total production.

THE INDUSTRIAL TREATMENT OF COAL

The treatment of coal, in order to obtain coal gas and the raw materials from which fertilizers and other chemicals are derived, involves heating the coal in the absence of air at a high temperature to separate its components. If a hard coal such as anthracite was heated in this way, the main product would be coke. A soft coal, on the other hand, would be almost completely volatilized and would yield very little coke. The actual method of treatment, then, depends on the nature of the coal and the products required.

The Manufacture of Coke and Coal Gas

Hard coal (bituminous + anthracite) is heated at about 1300°C in vertical or horizontal iron retorts. The coal is decomposed into crude coal gas and coke (impure carbon). The gas is forced out of the retorts by water gas which is produced by passing steam in at the base of the retorts. Coke comes out of the bottom of the retorts and is quenched in cold water.

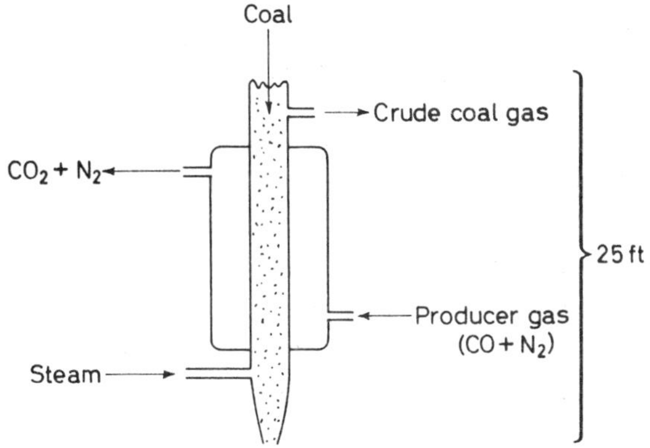

Figure 3.2

The crude coal gas is cooled and allowed to separate into (i) gas, (ii) ammoniacal liquor, and (iii) coal tar. The latter two are run off into a large separating tank and the gas passes on for further treatment.

At this stage the gas contains ammonia, hydrogen sulphide, aromatic vapours and small amounts of other substances which must be removed before the gas is suitable for use as a domestic fuel. Last traces of ammonia are removed from the gas by washing with water, and hydrogen sulphide is removed by passing the gas over heated iron oxide.

$$Fe_2O_3 + 3H_2S = Fe_2S_3 + 3H_2O$$

The oxide is regenerated by heating in air.

$$2Fe_2S_3 + 9O_2 = 2Fe_2O_3 + 6SO_2$$

Sulphur dioxide, produced in the regeneration process, is used to make sulphuric acid.

Aromatic substances which remain at this stage are removed by washing the gas with oil. The aromatics are recovered from the oily mixture by distillation.

The final composition of coal gas is approximately as follows:

Hydrogen	50%
Methane	35%
Carbon dioxide	6%
Other gases	9%

Coal gas is sold on the basis of so much per *therm*, where a therm is the amount of heat required to raise the temperature of a 100,000 lb. of water through 1°F (i.e. 100,000 British Thermal Units). The heating value of coal gas is determined by the gas boards and is stated on their bills as so many British Thermal Units (B.t.u.) per cubic foot.

Coal tar contains numerous organic substances and may be partially separated by distillation. Primary distillation separates the tar into five fractions (*Table 3.2*). The chemicals constituting the various fractions (see *Table 3.3*) are starting materials in the manufacture of many important substances. Adipic acid (used to make Nylon) and medicines such as aspirin (see p. 115) and phenacetin

Table 3.2

Fraction	*Boiling range,* °C	*Main constituents*
Light oil	up to 170	Benzene, toluene, xylene
Middle oil	170–230	Naphthalene, phenol
Heavy oil	230–270	Naphthalene, naphthol
Green oil	270–400	Anthracene
Pitch	above 400	

Table 3.3

Substance	*Formula*	*Structure*
Benzene	C_6H_6	
Toluene	$C_6H_5CH_3$	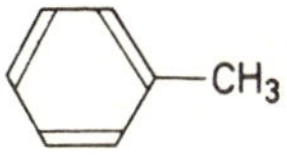 CH3
Xylenes	$C_6H_4(CH_3)_2$	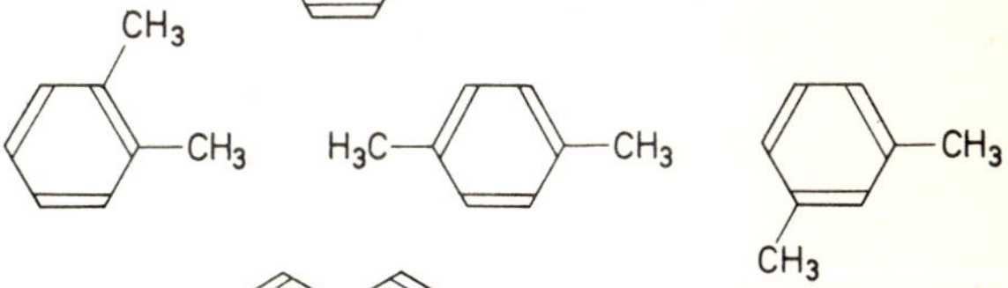 CH3 CH3 H3C CH3 CH3 CH3
Naphthalene	$C_{10}H_8$	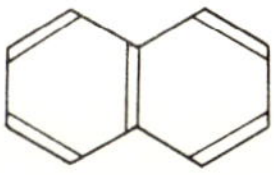
Phenol	C_6H_5OH	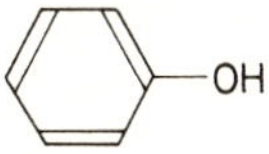 OH
Naphthol	$C_{10}H_7OH$	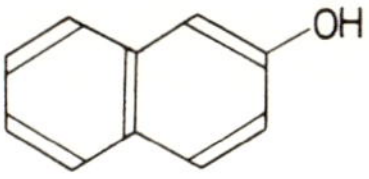 OH
Anthracene	$C_{14}H_{10}$	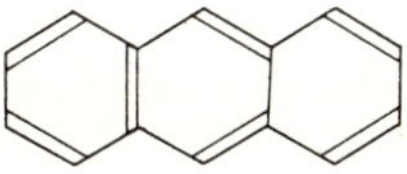
Phenanthrene	$C_{14}H_{10}$	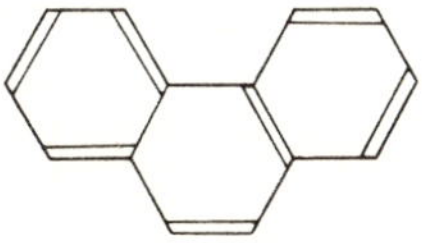

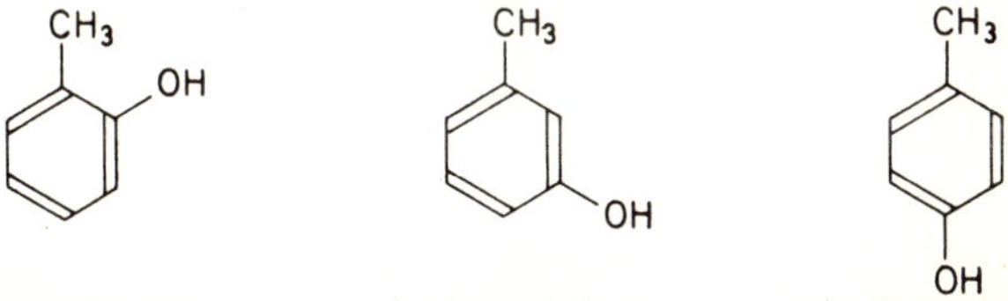

CH3 OH CH3 OH CH3 OH

o-hydroxy toluene *m*-hydroxy toluene *p*-hydroxy toluene

Cresols

are derived from phenols. The antiseptic T.C.P. is obtained from cresols (hydroxytoluenes). The disinfectant, Dettol, is obtained from xylenols [hydroxy xylenes, $(CH_3)_2C_6H_3OH$].

Naphthalene and anthracene are starting materials for the manufacture of dyes (e.g., indigo and alizarin).

The Lurgi Process

Lower-grade coals containing a high proportion of ash and volatile matter are not suitable for making coke. These coals can be treated in a modern process, the Lurgi process, which originated in Germany. In this case the coal is completely gasified by reacting with steam and oxygen at about 30 atmospheres (see *Figure 3.3*).

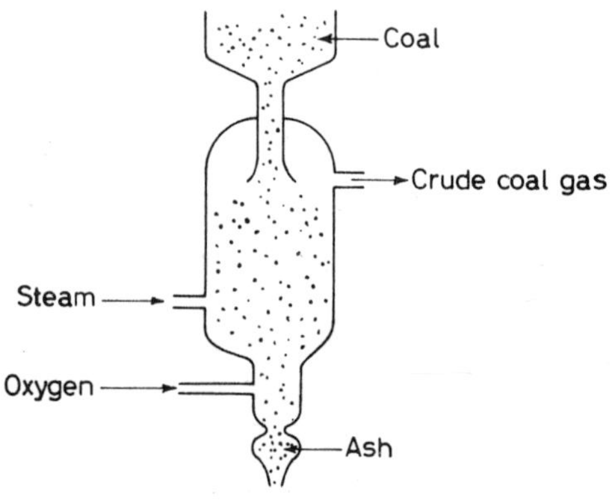

Figure 3.3

The crude product is fractionated as before to obtain benzene and benzene aromatics, and the crude coal gas is enriched with hydrogen. This is necessary because the calorific value of the gas made by the Lurgi process is lower than that made by conventional methods. Carbon monoxide in the coal gas is converted to carbon dioxide by reaction with steam in the presence of a catalyst.

$$CO + H_2O \xrightarrow{\text{catalyst}} H_2 + CO_2$$

By this means carbon monoxide is replaced by the more calorific hydrogen.

EXPERIMENTS

Set up the apparatus as shown in *Figure 3.4* and heat the coal until no further decomposition occurs.

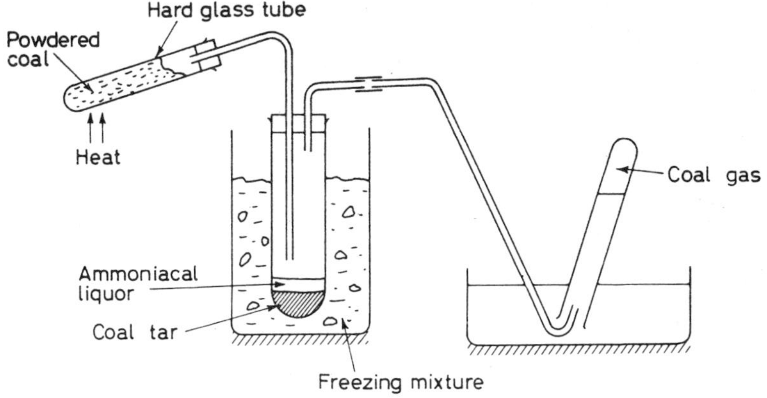

Figure 3.4

Tests

(a) On the gas

(i) Ignite a test-tube of the gas. What is the colour of the flame?

(ii) Shake the gas with a little dilute alkaline potassium permanganate solution. Describe and explain what happens.

(iii) Partially fill a test-tube with coal gas, in the manner indicated above, then allow the remaining water to run out and air to take its place. Ignite the gas in the tube and describe what happens. Repeat a number of times using different proportions of air and coal gas.

What would you incorporate into the apparatus to show the presence of hydrogen sulphide in the coal gas?

(b) On the ammoniacal liquor

(iv) Test the liquid with litmus paper. Is it acid or alkaline?

(v) Add a few ml. of dilute sodium hydroxide solution, warm the mixture and detect any gas evolved.

QUESTIONS

1. Write a short essay about coal.
2. Explain how coal was formed. Describe the process by which coal gas is obtained from coal.
3. 'Coal is more important than petroleum.' Do you agree with this statement? Give reasons for your answer.
4. What chemical reactions are involved in the manufacture of coal gas from coal? Give equations.
5. What is the composition of coal? Discuss the most important uses of coal.

6. How would you attempt to determine the proportions of hydrogen, methane and carbon dioxide in coal gas?
7. Describe an experiment to identify the chief products of the combustion of coal gas. What is the approximate composition of coal gas?
8. What is meant by the term, hydrocarbon? Give suitable examples. Outline briefly how benzene is obtained from coal.
9. A kettle holds 3 pints of water (1 pint = 550 g) and it is required to raise the temperature of the water from 15°C to 100°C by heating on a gas stove. If 1 therm is equivalent to 200 cu. ft. of gas. what volume of gas is required to boil the water in the kettle? (Neglect heat capacity of the kettle.) 1 B.t.u. = 251 calories.
10. What is the 'carbon cycle'?

PART II

DERIVATIVES OF HYDROCARBON COMPOUNDS

MANY simple organic compounds contain, besides carbon and hydrogen, one or more of the following elements: oxygen, nitrogen, halogens, sulphur, phosphorus, arsenic and metals such as iron, cobalt and magnesium. Organic compounds containing these other elements may be considered to be derivatives of hydrocarbons. A few of the simpler and more familiar of these hydrocarbon derivatives are given, along with their formulae, in *Table 3.4*.

Table 3.4

Substance	*Formula*
Ethanol	C_2H_5OH
Methanol	CH_3OH
Formic acid	$HCOOH$
Acetic acid	CH_3COOH
Diethyl ether	$C_2H_5OC_2H_5$
Methyl ethyl ether	$CH_3OC_2H_5$
Acetaldehyde	CH_3CHO
Acetone	CH_3COCH_3
Aniline	$C_6H_5NH_2$
Methyl chloride	CH_3Cl
Iodobenzene	C_6H_5I
Ethyl hydrogen sulphate	$C_2H_5HSO_4$
Thiophene	S

In the chapters which follow, some of these compounds and their homologues are considered in more detail.

4

ALCOHOLS

INTRODUCTION

ALCOHOLS are organic compounds containing a hydroxyl group. Simple alcohols contain just one hydroxyl group and are represented by the general formula, ROH, where R has its usual significance denoting an alkyl group.

The simple alcohols form a homologous series of compounds and the first member of the series is methanol.

Table 4.1

Common name	*Modern name*	*Formula*
Methyl alcohol	Methanol	CH_3OH
Ethyl alcohol	Ethanol	C_2H_5OH
n-Propyl alcohol	Propan-1-ol	$CH_3CH_2CH_2OH$
n-Butyl alcohol	Butan-1-ol	$CH_3CH_2CH_2CH_2OH$

The modern, systematic method of naming the alcohols has been devised to overcome the problem of differentiating between alcohols having the same molecular formula but different structural formulae. In the case of methanol there is no difficulty since there is only one alcohol having the molecular formula CH_4O. Similarly with ethanol, but with propanol, C_3H_8O, there are *two* alcohols having this formula

i.e. $CH_3CH_2CH_2OH$ propan-1-ol

$$\underset{\displaystyle\overset{|}{OH}}{CH_3CH{\cdot}CH_3}$$ propan-2-ol

For the sake of naming these substances they are considered to be derived from the parent hydrocarbon, propane. The number in the middle of the name denotes the position of the carbon atom to which the hydroxyl group is attached. The ending, -ol, indicates an alcohol.

Propan-1-ol is known as a primary alcohol because it contains the group $—CH_2OH$. Propan-2-ol is a secondary alcohol since it contains the group $>CHOH$. Tertiary alcohols contain the group $\geq COH$.

There are *four* butanols.

$$CH_3CH_2CH_2CH_2OH$$

butan-1-ol

$$\begin{array}{l} CH_3 \\ \quad \diagdown \\ \quad\quad CHCH_2OH \\ \quad \diagup \\ CH_3 \end{array}$$

2 methylpropan-1-ol
(*iso*butanol)

$$\begin{array}{l} CH_3CH_2 \\ \quad\quad \diagdown \\ \quad\quad\quad CHOH \\ \quad\quad \diagup \\ CH_3 \end{array}$$

butan-2-ol

$$\begin{array}{c} CH_3 \\ | \\ CH_3—C—OH \\ | \\ CH_3 \end{array}$$

2-methylpropan-2-ol

METHODS OF PREPARATION

The best-known alcohols are methanol and ethanol. They are used extensively as solvents for paints, varnishes and gums and for the manufacture of dyes, perfumes, resins and other solvents.

Methanol is made industrially by the destructive distillation of wood and from water gas. Only a small percentage of the total production comes from wood (the alcohol is a by-product in the manufacture of wood charcoal) but it was this method which gave rise to the name, 'wood spirit' for methanol. In the modern synthesis water gas is mixed with half its volume of hydrogen and the mixture passed at a pressure of 200 atmospheres over a catalyst of zinc and chromium oxides at 300°C.

$$CO + 2H_2 = CH_3OH$$

Pure ethanol for industrial use is made from ethene.

(a) In the most recent method ethene is treated with steam under pressure and in the presence of a catalyst.

$$CH_2=CH_2 + H_2O \xrightarrow[H_3PO_4/\text{Kieselguhr}]{300°C,\ 1000\ \text{p.s.i.}} C_2H_5OH$$

(b) In an older method ethene is treated with sulphuric acid and the product is hydrolysed to ethanol.

REACTIONS

$$CH_2{=}CH_2 + \underset{(H_2SO_4)}{HOSO_2OH} \rightarrow \underset{\text{ethyl hydrogen sulphate}}{CH_3CH_2OSO_2OH}$$

$$\downarrow + H_2O$$

$$C_2H_5OH + HOSO_2OH$$

Ethanol is still obtained by fermentation but this process is mostly used in making beer and wines.

In the brewing of beer, starch, obtained from potatoes and wheat, is mixed with malt (sprouted barley) which contains the enzyme diastase, and starch is converted to maltose (a sugar).

$$n\,H_2O + 2\,[C_6H_{10}O_5]_n \xrightarrow[\text{1 hour}]{50°C} n\,C_{12}H_{22}O_{11}$$
$$\text{Diastase}$$

The liquor is further fermented with yeast which contains the enzymes, zymase and maltase, when maltose is converted to ethanol.

$$\underset{\text{maltose}}{C_{12}H_{22}O_{11}} + H_2O \xrightarrow[30°C]{\text{Maltase}} 2\,\underset{\text{glucose}}{C_6H_{12}O_6}$$

$$C_6H_{12}O_6 \xrightarrow[30°C]{\text{Zymase}} 2\,\underset{\text{ethanol}}{C_2H_5OH} + 2\,CO_2$$

Conversion of maltose to ethanol takes between 1 and 3 days. The final liquor contains 6–10% alcohol.

Carbon dioxide formed in the reaction is sold as a byproduct, i.e., as 'Dry ice' (solid carbon dioxide).

The fermentation processes for making alcohols, and also lactic acid, hormones and penicillins (see p. 119), have the obvious advantages of low-temperature operation, mild conditions reducing the probability of corrosion and high yields. The disadvantages are that the processes are slow and require experts to prepare and handle the enzymes.

PROPERTIES

The lower members of the alcohol series are volatile, neutral liquids. They have a distinctive smell and taste and are soluble in water. The higher members are only partially soluble, or insoluble, in water and are viscous liquids or solids at room temperature.

REACTIONS

The hydroxyl group (–OH) is called a *functional* group because, as we shall see later, it is this group which is involved in the simple reactions of alcohols and hence decides the chemical properties of the alcohols.

(i) The alcohols react with metallic sodium to yield hydrogen and form substances known as alkoxides e.g. $$2CH_3OH + 2Na \rightarrow \underset{\text{sodium methoxide}}{2CH_3ONa} + \overrightarrow{H_2}$$

$$\underset{\text{propan-1-ol}}{2CH_3CH_2CH_2OH} + 2Na \rightarrow \underset{\text{sodium propoxide}}{2CH_3CH_2CH_2ONa} + \overrightarrow{H_2}$$

(ii) With phosphorus halides the alcohols form alkyl halides (see p. 65)

e.g. $$\underset{\text{ethanol}}{C_2H_5OH} + PCl_5 \rightarrow \underset{\text{ethyl chloride}}{C_2H_5Cl} + POCl_3 + \overrightarrow{HCl}$$

$$\underset{\text{butan-1-ol}}{CH_3CH_2CH_2CH_2OH} + PCl_5 \rightarrow \underset{\text{1-chlorobutane}}{CH_3CH_2CH_2CH_2Cl} + POCl_3 + \overrightarrow{HCl}$$

(iii) Alcohols react with acids to form esters (see p. 45)

e.g. $$\underset{\text{acetic acid}}{CH_3COOH} + C_2H_5OH \xrightarrow[H_2SO_4]{\text{conc.}} \underset{\text{ethyl acetate}}{CH_3COOC_2H_5} + H_2O$$

(iv) Alcohols can be converted to aldehydes or ketones (see p. 57) by the action of oxidizing agents. This is an important reaction since it can be used as a means of distinguishing between primary and secondary alcohols. Primary alcohols on oxidation yield aldehydes whilst secondary alcohols yield ketones

e.g. $$\underset{\text{propan-1-ol}}{CH_3CH_2CH_2OH} + [O] \xrightarrow[H_2SO_4]{K_2Cr_2O_7} \underset{\text{propanal}}{CH_3CH_2C(=O)H} + H_2O$$

$$\underset{\text{propan-2-ol}}{CH_3CH(OH)CH_3} + [O] \xrightarrow[H_2SO_4]{K_2Cr_2O_7} \underset{\text{acetone}}{CH_3C(=O){\cdot}CH_3} + H_2O$$

If the oxidizing agent is in excess, primary alcohols are oxidized to the corresponding acid

e.g. $$\underset{\text{propan-1-ol}}{CH_3CH_2CH_2OH} \xrightarrow{[O]} \underset{\text{propanal}}{C_2H_5CHO} \xrightarrow{[O]} \underset{\text{propanoic acid}}{C_2H_5COOH}$$

EXPERIMENTS

Preparation of Ethanol: Fermentation of Sugar

Grind 1 g of brewers yeast in a mortar and make into a paste with a little water. Add this paste to 10 g of sucrose and a few crystals of ammonium phosphate and potassium nitrate dissolved in a 100 g of water. Put the mixture in a 250 ml. flask fitted with a delivery tube (as shown in *Figure 4.1*) and keep in a warm place until no further carbon dioxide is evolved (5–7 days).

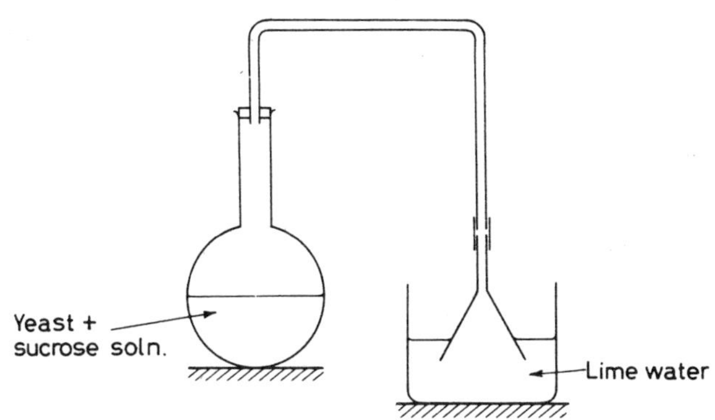

Figure 4.1

Filter the mixture and distil the filtrate (use the apparatus illustrated in *Figure 2.3* collecting the liquid which comes over between 70° and 95°C. Dry the fraction by allowing it to stand over anhydrous sodium sulphate. Filter and distil using a fractionating column (*Figure 2.3*) collecting the fraction coming over between 77° and 79°C.

2. Dry a *small* piece of sodium on a filter paper and transfer it to a test-tube containing about 5 ml. of methanol. Notice what happens and when the reaction has finished, pour the solution into an evaporating basin and evaporate to dryness on a water bath. What is the solid remaining?

3. Wrap a piece of copper gauze round a stout piece of copper wire (about 6 in. long) bent at both ends to form a hook. Heat the gauze

in a hot bunsen flame until a coating of black oxide is formed on its surface. Whilst the gauze is still hot (not red hot!), plunge it into about 5 ml. of methanol contained in a boiling-tube. Notice the reduction of black copper oxide to copper and also the penetrating odour. Explain what happens to the alcohol.

4. Put five drops of ethanol in a test-tube. Add five drops of glacial acetic acid (take care not to inhale the fumes) and five drops of concentrated sulphuric acid (add this very carefully). Warm the mixture gently and detect any change in odour. Write the equation for the reaction.

5. Add 1 ml. of concentrated sulphuric acid to 2 ml. of potassium dichromate solution. Shake the mixture and cool under the cold tap. Add 1 ml. of ethanol and notice the colour changes and any new odours. Write simplified equations for the reactions which have occurred involving ethanol.

QUESTIONS

1. What is meant by the term fermentation? Discuss briefly the advantages and disadvantages of fermentation processes.
2. What is an alcohol? Name three alcohols and write their structural formulae.
3. If you were presented with a colourless, neutral liquid which had a burning taste, and you suspected it to be an alcohol, what chemical tests would you perform to confirm your suspicion? Write equations.
4. How may ethene be converted to ethanol?
5. How may carbon monoxide be converted to methanol?
6. Write the structural formulae for the alcohols having the molecular formula $C_4H_{10}O$. Name the alcohols.
7. Explain how ethanol is made industrially. Write the equations for the reactions of ethanol with (a) Na, (b) PCl_5, and (c) $K_2Cr_2O_7 + H_2SO_4$. Name the products.
8. Describe how ethanol is made from starch. What are the characteristic reactions of the hydroxyl group?
9. Suppose you suspect a colourless organic liquid to be ethanol, what tests would you perform to prove its identity?
10. How may ethanol be distinguished from methanol? (Hint: Reactions with iodine.)

5

ORGANIC ACIDS AND ESTERS

INTRODUCTION

ORGANIC acids have the general formula, RCOOH. The group R represents an *alkyl* group. This may be a simple methyl (CH_3–) or ethyl (C_2H_5–) group or something more complex like $C_{15}H_{31}$— which is the alkyl group of palmitic acid.

The —COOH group is called a carboxyl group and it is this group of atoms which gives the molecule its acid properties in aqueous solutions

i.e. $$RCOOH + H_2O \rightleftharpoons RCOO^- + H_3O^+$$

In the case of an organic acid where R is a simple alkyl group the tendency for dissociation in water, as shown in the equation above, is relatively slight and the resulting solution is only weakly acidic.

Compounds containing one carboxyl group (monocarboxylic acids) form a homologous series. Members of the series have similar chemical properties.

Table 5.1

Common name	*Formula*	*Modern name*
Formic acid	$HCOOH$	Methanoic acid
Acetic acid	CH_3COOH	Ethanoic acid
Propionic acid	C_2H_5COOH	Propanoic acid
n-Butyric acid	C_3H_7COOH	Butanoic acid
n-Valeric acid	C_4H_9COOH	Pentanoic acid

In *Table 5.1*, the ending -oic in the modern names indicates the presence, in the compound, of a carboxyl group. Formic acid is the simplest monocarboxylic acid but many of its properties are not representative of the series. Acetic acid is the first typical member.

PREPARATION OF ORGANIC ACIDS

Formic Acid

Formic acid is obtained commercially by heating sodium hydroxide with carbon monoxide. Sodium hydroxide is powdered and heated to about 150°C; carbon monoxide, at about 10 atmospheres pressure, is passed over the salt.

$$\underset{}{NaOH + CO} \xrightarrow[150°C]{10\,atm} \underset{\text{sodium formate}}{HCOONa}$$

The free acid is obtained from sodium formate by distillation with dilute sulphuric acid.

$$HCOONa + H_2SO_4 \rightarrow \underset{\text{formic acid}}{HCOOH} + NaHSO_4$$

In the laboratory the acid can be made by heating oxalic acid with glycerol at about 120°C.

$$\underset{\text{oxalic acid}}{\begin{array}{c} COOH \\ | \\ COOH \end{array}} \xrightarrow[120°C]{\text{Glycerol}} HCOOH + CO_2$$

Acetic Acid

Acetic acid is made industrially by oxidizing ethanol or acetaldehyde (CH_3CHO). Alcohol or aldehyde vapour mixed with air is passed over a heated metal catalyst.

$$\underset{\text{ethanol}}{CH_3CH_2OH} + O_2 \xrightarrow{\text{catalyst}} \underset{\text{acetic acid}}{CH_3COOH} + H_2O$$

$$\underset{\text{acetaldehyde}}{2CH_3CHO} + O_2 \xrightarrow{\text{catalyst}} \underset{\text{acetic acid}}{2CH_3COOH}$$

In the laboratory it is more convenient to use potasium dichromate as the oxidizing agent and perform the reaction in the liquid phase (*Figure 5.1*).

$$CH_3CH_2OH + 2[O] \rightarrow CH_3COOH + H_2O$$

(oxygen from $K_2Cr_2O_7 + H_2SO_4$)

REACTIONS OF ORGANIC ACIDS

(1) Organic acids, like inorganic acids, will react with strong bases forming salts. Acetic acid reacts with sodium hydroxide to form sodium acetate.

$$CH_3COOH + NaOH \rightarrow CH_3COONa + H_2O$$

In this reaction the hydrogen atom of the carboxyl group is replaced by sodium.

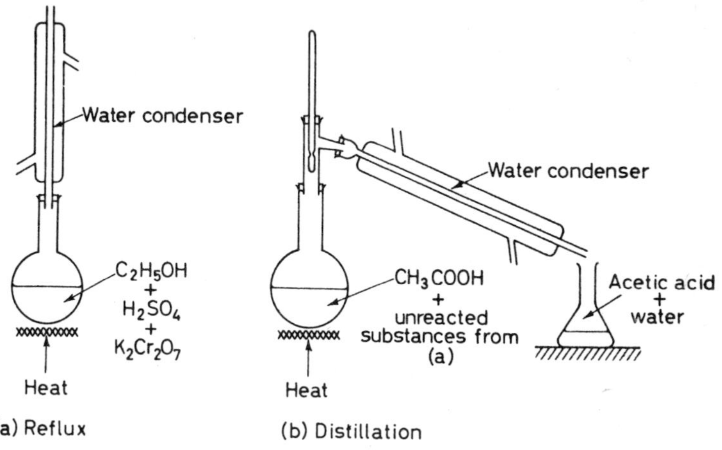

Figure 5.1. *(a) Reflux; (b) distillation*

(2) With phosphorus chlorides the acids react forming *acid chlorides* (also called acyl chlorides).

$$\underset{}{CH_3COOH} + \underset{\substack{\text{phosphorus}\\\text{pentachloride}}}{PCl_5} \rightarrow \underset{\text{acetyl chloride}}{CH_3COCl} + POCl_3 + HCl$$

Formic acid, unlike other organic acids, does not form an acid chloride.

(3) The acids may be dehydrated to compounds called *anhydrides*. For example, when acetic acid is boiled with sufficient phosphorus pentoxide (P_2O_5) it is converted to acetic anhydride.

$$\begin{matrix} CH_3COOH \\ CH_3COOH \end{matrix} \xrightarrow[P_2O_5]{-H_2O} \begin{matrix} CH_3C(=O) \\ \quad \diagdown O \\ \quad \diagup \\ CH_3C(=O) \end{matrix}$$

acetic anhydride

(4) Organic acids will react with alcohols, in the presence of a little concentrated sulphuric acid, to produce substances collectively called *esters*. Acetic acid reacts with ethanol to form ethyl acetate.

$$CH_3COOH + C_2H_5OH \xrightarrow{H_2SO_4} \underset{\text{ethyl acetate}}{CH_3COOC_2H_5} + H_2O$$

In order that the reaction shall go to the right, sulphuric acid must be present; it acts as a catalyst and as dehydrating agent.

SYNTHESIS

The simple ester ethyl acetate may be used to illustrate organic synthesis. Synthesis is defined as the building up of small organic molecules into larger ones. In the following example ethyl acetate is synthesized from carbon.

Step 1

Carbon is first converted to calcium carbide by heating with calcium oxide at a high temperature in an electric furnace.

$$CaO + 3C \rightarrow CaC_2 + CO$$

Step 2

Calcium carbide is then treated with water to produce the gas ethyne.

$$CaC_2 + 2H_2O \rightarrow C_2H_2 + Ca(OH)_2$$

Step 3

Ethyne may be converted to (a) acetaldehyde, from which acetic acid can be obtained, and (b) ethene, from which ethanol is made.

(a)

$$C_2H_2 \xrightarrow[+HgSO_4/FeSO_4 \text{ at } 60°C]{\text{Dil } H_2SO_4} \underset{\text{acetaldehyde}}{CH_3CHO}$$

$$2CH_3CHO + O_2 \xrightarrow{\text{catalyst}} \underset{\text{acetic acid}}{2CH_3COOH}$$

(b)

$$C_2H_2 + H_2 \xrightarrow[\text{catalyst}]{Ni} \underset{\text{ethene}}{C_2H_4}$$

$$C_2H_4 \xrightarrow{H_2SO_4} \underset{\text{ethyl hydrogen sulphate}}{C_2H_5HSO_4}$$

$$C_2H_5HSO_4 + H_2O \rightarrow \underset{\text{ethanol}}{C_2H_5OH} + H_2SO_4$$

Step 4

The final stage in the synthesis consists of combining acetic acid and ethanol in the manner previously described.

$$CH_3COOH + C_2H_5OH \xrightarrow{H_2SO_4} \underset{\text{ethyl acetate}}{CH_3COOC_2H_5} + H_2O$$

USES OF ORGANIC ACIDS

Formic acid is used in the synthetic fibre industry as a solvent for Nylon. It is also used for coagulating rubber latex (p. 101) and in the dyeing industry.

Acetic acid is used for making cellulose acetate and polyvinyl acetate. Polyvinyl acetate is made by the following sequence of reactions.

$$\underset{\text{ethyne}}{CH{\equiv}CH} + CH_3COOH \xrightarrow[\text{HgSO}_4\ \text{catalyst}]{850^\circ C} CH_3COOCH{=}CH_2$$

$$\xrightarrow{+\ \text{benzoyl peroxide}} \underset{\text{polyvinyl acetate}}{\left[-CH_2-\underset{\substack{| \\ OCOCH_3}}{CH}-CH_2 \right]_n}$$

In addition, acetic acid is used as a preservative (in vinegar), as a solvent, bleaching agent and for making important derivatives

e.g. $$CH_3COOH + Cl_2 \rightarrow \underset{\text{monochloroacetic acid}}{CH_2ClCOOH} + HCl$$

$$CH_2ClCOOH + Cl_2 \rightarrow \underset{\text{dichloroacetic acid}}{CHCl_2COOH} + HCl$$

$$CHCl_2COOH + Cl_2 \rightarrow \underset{\text{trichloroacetic acid}}{CCl_3COOH} + HCl$$

$$CH_3COOH \xrightarrow[\text{heat}]{NH_3} \underset{\text{acetamide (ethanamide)}}{CH_3CONH_2}$$

USES OF ORGANIC ESTERS

Esters find a very wide application. They often have a pleasant smell and for this reason are used in the perfumery industry.

Many synthetic fibres are esters, e.g. cellulose acetate, Terylene, polyvinyl acetate and Perspex. Naturally-occurring fats and waxes are esters (see soap below).

Soap

Soap consists of the salts of high-molecular-weight organic acids. Ordinary soap contains the sodium salts of stearic and palmitic acids.

$$\underset{\text{sodium stearate}}{C_{17}H_{35}COONa} \qquad \underset{\text{sodium palmitate}}{C_{15}H_{31}COONa}$$

Soap is made by heating fats (Chapter 10) with sodium hydroxide.

In this reaction the glycerol ester breaks up into glycerol and the sodium salts of fatty (carboxylic acids) acids.

$$\begin{array}{l} CH_2OOC{\cdot}C_{17}H_{35} \\ | \\ CHOOC{\cdot}C_{17}H_{35} \\ | \\ CH_2OOC{\cdot}C_{17}H_{35} \end{array} + 3NaOH \rightarrow \begin{array}{l} CH_2OH \\ | \\ CHOH \\ | \\ CH_2OH \end{array} + 3C_{17}H_{35}COONa$$

glyceryl tristearate (fat) — glycerol — sodium stearate

The process is known as *saponification* and in a little more detail involves the following stages. Sodium hydroxide is added to fat in a large vat. Steam is passed through the mixture and when saponification has occurred sodium chloride is added to salt out the soap. This floats on the surface of the mixture and is scooped off. The aqueous solution, which contains glycerol, is sucked off and distilled to obtain glycerol.

EXPERIMENTS

1. Test the solubility of some common organic acids in water.

Acetic acid	(CH_3COOH)
Oxalic acid	$(COOH)_2$
Monochloroacetic acid	($ClCH_2COOH$)
Succinic acid	$(CH_2COOH)_2$

2. Test the solubility of these acids in dilute alkali (e.g. sodium hydroxide solution).

3. Discover how these acids react with sodium carbonate solution. Place 5 ml. of sodium carbonate solution in a test-tube. Add a piece of porous pot and boil for a few minutes to ensure that no bicarbonate or free carbon dioxide is present. Cool, and add about 0·1 g of oxalic acid. Repeat for other acids.

4. Organic acids give characteristic colorations with ferric chloride. For the test to be successful both acid and chloride must be made neutral. Proceed as follows.

(i) Place 1 ml. of acid in a boiling tube and add a slight excess of ammonia solution until just alkaline to litmus. Add a piece of porous pot to the tube and boil solution until the odour of ammonia is completely removed.

(ii) Ferric chloride solution is usually acid. Make alkaline by adding dilute sodium hydroxide solution until a precipitate of ferric hydroxide is *just* formed, filter off the precipitate and use the filtrate.

Add a little of (i) to (ii). Test a number of different organic acids and record the colour changes.

5. Heat the following acids with soda lime:
   (i) benzoic acid
   (ii) phthalic acid
   (iii) sodium acetate

Take about 0·5 g of acid and 2 g of soda-lime in each case. Notice the smell and explain what is occurring.

Try burning any gas which is evolved.

6. To 1 ml. of pentan-1-ol add 1 ml. of glacial acetic acid and then add *dropwise* 0·5 ml. of concentrated sulphuric acid. Warm the mixture in hot water and then pour into a 100 ml. of cold water contained in a beaker. Notice the odour (an ester). Write the equation for the reaction.

QUESTIONS

1. Write the names and formulae of the first five members of the aliphatic monocarboxylic acid homologous series.
2. Describe the laboratory preparation of acetic acid. Draw a diagram of the apparatus used and write an equation for the reaction.
3. Acetic acid is a weak acid. Explain this statement.
4. Why is formic acid not typical of the acid homologous series? Explain by reference to the chemical properties of the acid.
5. Write equations for the reaction of acetic acid with (i) sodium carbonate, (ii) phosphorus pentachloride, (iii) phosphorus pentoxide, and (iv) ethanol.
6. Explain the term *synthesis*.
7. Outline the uses of organic acids.
8. How may ethyl acetate be prepared? Discover and describe the characteristic odours of ethyl acetate, amyl acetate, methyl salicylate, ethyl butyrate and ethyl formate.
9. Describe the manufacture of soap.
10. Describe the manufacture of Terylene (see Chapter 11).

6

AMINES

INTRODUCTION

AMINES are organic compounds containing nitrogen. They are important since their chemical properties are indicative of the $-NH_2$ group which occurs, for example, in amino-acids (p. 86).

Amines form a homologous series in which the $-NH_2$ group is the functional group. They may be considered to be derived from ammonia by the replacement of hydrogen atoms in ammonia by alkyl groups.

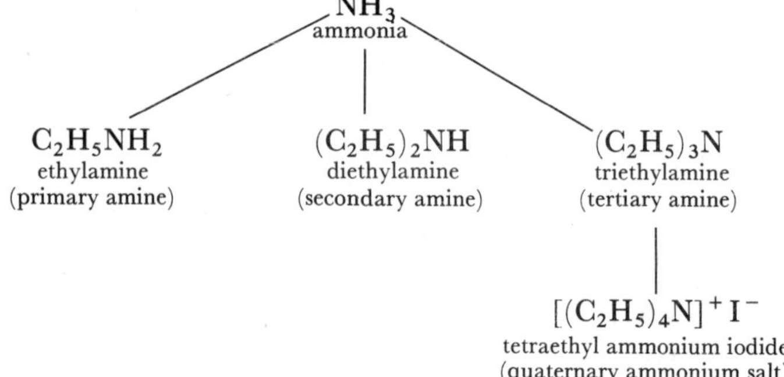

PREPARATION OF AMINES

(1) The most convenient laboratory preparation of primary amines consists of heating an amide (i.e. substance containing the $-CONH_2$ group) with bromine and alkali. This method is known as Hofmann's method

e.g. $$\underset{\text{propanamide}}{C_2H_5CONH_2} + Br_2 + 4KOH \rightarrow \underset{\text{ethylamine}}{C_2H_5NH_2} + K_2CO_3 + 2KBr + 2H_2O$$

For convenience the amine is collected in dilute hydrochloric acid, in which it dissolves forming a salt. The salt is isolated by evaporating the solution and recrystallizing from absolute alcohol.

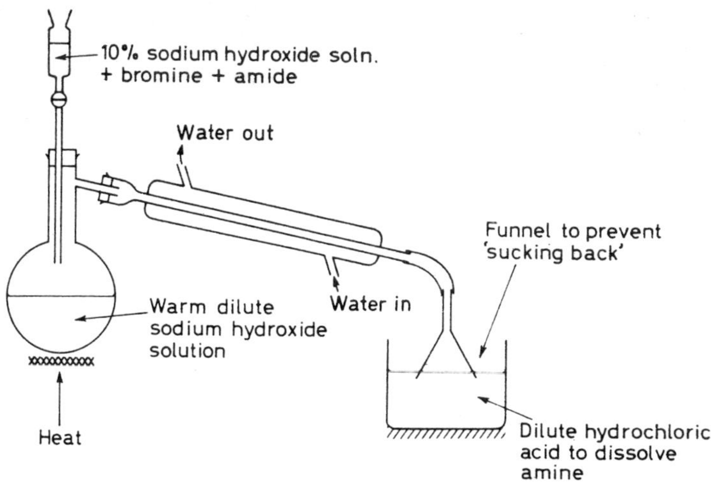

Figure 6.1

$$C_2H_5NH_2 + HCl \rightarrow \underset{\text{ethylamine hydrochloride}}{C_2H_5NH_3^+Cl^-}$$

If the free base is required it may be obtained from the salt by warming the latter with 20% sodium hydroxide solution. Ethylamine comes off as a vapour and is collected in a cooled receiver.

$$C_2H_5NH_3^+Cl^- + NaOH \rightarrow C_2H_5NH_2\uparrow + NaCl + H_2O$$

(2) Amines can be made by treating alkyl halides with ammonia in alcohol.

$$\underset{\text{ethyl iodide}}{C_2H_5I} + NH_3 \text{ (alcohol)} \rightarrow \underset{\text{ethylamine}}{C_2H_5NH_2} + HI$$

In practice the reaction is not as simple as indicated above since ethylamine can react with more ethyl iodide to give a number of products.

$$\underset{\text{ethylamine}}{C_2H_5NH_2} + \underset{\text{ethyl iodide}}{C_2H_5I} \rightarrow \underset{\text{diethylamine}}{(C_2H_5)_2NH} + HI$$

$$\underset{\text{diethylamine}}{(C_2H_5)_2NH} + C_2H_5I \rightarrow \underset{\text{triethylamine}}{(C_2H_5)_3N} + HI$$

$$\underset{\text{triethylamine}}{(C_2H_5)_3N} + C_2H_5I \rightarrow \underset{\text{tetraethylammonium iodide}}{[(C_2H_5)_4N]^+I^-}$$

The final product is known as a quaternary ammonium salt and in this case is tetraethylammonium iodide. Quaternary ammonium salts may be thought of as ammonium compounds in which the hydrogen has been replaced by alkyl groups

e.g. NH_4I (ammonium iodide) $[(C_2H_5)_4N]^+I^-$ (tetraethyl ammonium iodide)

Amines are prepared on an industrial scale by the method described above but it is not suitable as a laboratory method since it is difficult to separate the products. A partial separation may be effected by fractional distillation.

Other methods of manufacture of amines include, heating the appropriate alcohol with ammonia under pressure and in the presence of a catalyst (aluminium oxide or cobalt), and reduction of nitro-paraffins.

PROPERTIES OF AMINES

1. They have an odour like ammonia but less pungent and more fish-like.
2. Their vapour turns red litmus blue (basic).
3. They dissolve in water to give alkaline solutions and they react with acids to form salts.

$$\underset{\text{ethylamine}}{C_2H_5NH_2} + HCl \rightarrow \underset{\text{ethylamine hydrochloride}}{C_2H_5N^+H_3Cl^-}$$

4. They burn in oxygen—the lower members with a blue flame.
5. Amines react with nitrous acid to form an alcohol and nitrogen.

$$\underset{\text{ethylamine}}{C_2H_5NH_2} + HNO_2 \rightarrow \underset{\text{ethanol}}{C_2H_5OH} + N_2\uparrow + H_2O$$

$$\underset{\text{aniline}}{C_6H_5NH_2} + HNO_2 \rightarrow \underset{\text{phenol}}{C_6H_5OH} + N_2\uparrow + H_2O$$

At low temperatures aromatic amines react differently with nitrous acid. They form what are called *diazo*-compounds.

$$\underset{\text{Aniline}}{C_6H_5NH_2} + HNO_2 \xrightarrow[HCl]{<5^\circ C} \underset{\text{Benzene diazonium chloride}}{[C_6H_5N_2]^+Cl^-} + 2H_2O$$

These diazo compounds are important since they are used to make dyes (p. 108).

6. Amines react with acyl chlorides to form *N*-acyl compounds. This reaction is important as a means of characterizing amines.

$$\underset{\text{aniline}}{C_6H_5NH_2} + \underset{\text{acetyl chloride}}{CH_3COCl} \rightarrow \underset{\text{acetanilide}}{C_6H_5NHCOCH_3} + HCl$$

USES OF AMINES

Some high-molecular-weight amines are used to make plastics. Hexamethylene diamine for example ($NH_2(CH_2)_6NH_2$), is used to make Nylon (p. 97).

Quaternary ammonium groups are components of strong anion-exchange resins. These are substances used for removing chloride, sulphate, carbonate and bicarbonate ions from water, e.g. hydroxyl ions exchanged for chloride ions:

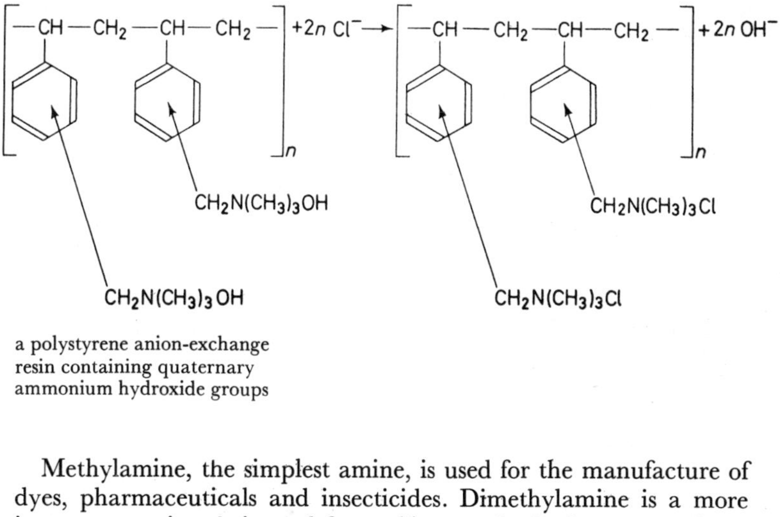

$[—CH—CH_2—CH—CH_2—]_n$ (each CH bearing a benzene ring with $CH_2N(CH_3)_3OH$) $+ 2n\ Cl^- \rightarrow$ $[—CH—CH_2—CH—CH_2—]_n$ (each CH bearing a benzene ring with $CH_2N(CH_3)_3Cl$) $+ 2n\ OH^-$

a polystyrene anion-exchange resin containing quaternary ammonium hydroxide groups

Methylamine, the simplest amine, is used for the manufacture of dyes, pharmaceuticals and insecticides. Dimethylamine is a more important amine, it is used for making synthetic detergents, drugs and dimethylformamide. The latter is used as a synthetic fibre solvent. Trimethylamine is used for making amine salts for adding to poultry feed to stimulate growth.

EXPERIMENTS

1. Smell a solution of methylamine and describe the odour.
2. Test whether a solution of methylamine is acidic or basic (use litmus).

3. Put 1 ml. of aniline in each of two test-tubes. To one tube add a few ml. of distilled water and shake. Notice that aniline is not soluble in water. Add dilute hydrochloric acid to the other test-tube and shake. Is aniline soluble in acid solution? Write an equation to explain what you think is happening.

4. Put 2 ml. of methylamine solution in a test-tube. Heat to boiling and hold the mouth of the tube close to a lighted taper. What happens? Could this test be used to distinguish between methylamine and ammonia? Explain.

5. Dissolve 5 drops of aniline in 3 ml. of dilute hydrochloric acid. Cool in ice water and then add 0·1 g of sodium nitrate dissolved in 1 ml. of water. The solution remains clear although a reaction has occurred and a diazonium chloride has been formed. (Save this solution and keep it cool.)

Dissolve 0·1 g of 2-naphthol in 1 ml. of dilute sodium hydroxide solution by warming gently, and then add 4 ml. of cold water. To this alkaline solution add the clear solution of the diazonium chloride from above. Describe what happens.

Note: This reaction only works for aromatic amines, i.e. those containing a benzene nucleus.

QUESTIONS

1. Give one example of each of the following: (i) a primary amine, (ii) a secondary amine, (iii) a tertiary amine, and (iv) a quarternary ammonium salt. Explain how they are related to ammonia.
2. Describe the laboratory preparation of methylamine.
3. Write equations for the reaction of ethylamine with (i) ethyl iodide, (ii) hydrochloric acid, (iii) nitrous acid, and (iv) acetyl chloride.
4. Enumerate the uses of amines.
5. What important naturally-occurring substances contain amine groups? Discover the chemical structure of these substances and find out how they may be tested for.
6. Write equations showing how it is possible to make the following conversions: (i) propanamide to ethylamine, (ii) ethylamine to tetraethyl ammonium iodide, (iii) ethylamine to ethyl chloride.

7

ALDEHYDES AND KETONES

INTRODUCTION

ALDEHYDES and ketones are considered together because they both contain what is called a carbonyl group, i.e. $>C=O$, and as a result they have similar chemical properties.

Aldehydes are identified by the –CHO group and ketones by the $>C=O$ group. Formaldehyde is the simplest aldehyde and acetone the simplest ketone.

$$H-C(=O)-H \qquad (CH_3)_2C=O$$

Formaldehyde Acetone

A few of the members of the aliphatic aldehyde and ketone homologous series are given in *Table 7.1*. In the modern names the ending -al indicates an aldehyde and the ending -one a ketone.

Table 7.1

Common name	*Modern name*	*Formula*
Aldehydes		
Formaldehyde	Methanal	$HCHO$
Acetaldehyde	Ethanal	CH_3CHO
Propionaldehyde	Propanal	CH_3CH_2CHO
Butyraldehyde	Butanal	$CH_3CH_2CH_2CHO$
Ketones		
Acetone	Propan-2-one	$(CH_3)_2CO$
Ethyl methyl ketone	Butan-2-one	$(CH_3)(C_2H_5)CO$
Diethyl ketone	Pentan-3-one	$(C_2H_5)_2CO$
Methyl propyl ketone	Pentan-2-one	$(C_3H_7)(CH_3)CO$

PREPARATION OF ALDEHYDES AND KETONES

Aldehydes and ketones can be made by oxidizing alcohols.

(a) If a primary alcohol (p. 39) is used, an aldehyde is obtained

e.g. $$\underset{\text{ethanol}}{C_2H_5OH} + [O] \rightarrow \underset{\text{acetaldehyde}}{CH_3CHO} + H_2O$$

The oxygen for this reaction comes from potassium dichromate. In acid solution, 1 gram molecule of potassium dichromate yields 3 gram atoms of oxygen for oxidation purposes

i.e. $$Cr_2O_7^{2-} + 8H^+ \rightarrow 2Cr^{3+} + 4H_2O + 3[O]$$

Note: The oxidizing agent must not be in excess otherwise the aldehyde may be oxidized to the corresponding acid.

$$\underset{\text{acetaldehyde}}{CH_3CHO} + [O] \rightarrow \underset{\text{acetic acid}}{CH_3COOH}$$

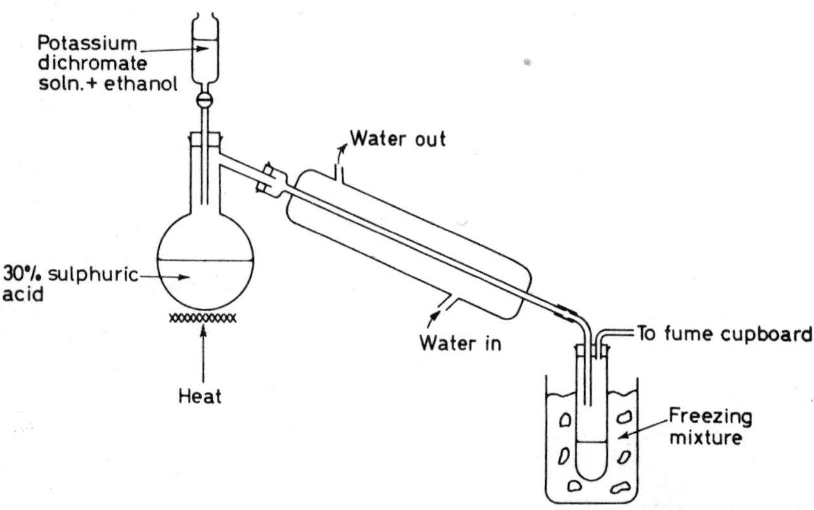

Figure 7.1. *Laboratory preparation of acetaldehyde*

(b) A ketone is obtained by oxidizing a secondary alcohol (p. 40).

$$\underset{\text{propan-2-ol}}{CH_3CH(OH)CH_3} + [O] \xrightarrow[H_2SO_4]{K_2Cr_2O_7} \underset{\text{acetone}}{CH_3\overset{\overset{\displaystyle O}{\|}}{C}{\cdot}CH_3} + H_2O$$

On a commercial scale, formaldehyde and acetaldehyde are made by oxidizing the appropriate alcohol with air in the presence of a granulated silver catalyst.

$$\underset{\text{methanol}}{CH_3OH} + \underset{\text{air}}{\tfrac{1}{2}O_2} \xrightarrow[400°C]{Ag} \underset{\text{formaldehyde}}{HCHO} + H_2O$$

$$\underset{\text{ethanol}}{CH_3CH_2OH} + \underset{\text{air}}{\tfrac{1}{2}O_2} \xrightarrow[450°C]{Ag} \underset{\text{acetaldehyde}}{CH_3CHO} + H_2O$$

Acetone is manufactured by the dehydrogenation of propan-2-ol in the presence of a zinc oxide catalyst at about 400°C.

$$CH_3CHOH{\cdot}CH_3 \xrightarrow[400°C]{ZnO} (CH_3)_2C{=}O + H_2$$

It is also obtained as a byproduct in the manufacture of phenol from benzene (Cumene process).

$$\text{Benzene} \xrightarrow[CH_3CH{=}CH_2]{AlCl_3} \underset{\text{Cumene}}{C_6H_5CH(CH_3)_2} \xrightarrow[120°C]{\text{Air}} \underset{\text{Hydroperoxide of cumene}}{C_6H_5C(CH_3)_2{-}O{-}OH} \xrightarrow[50°C]{\text{dil.}H_2SO_4} \underset{\text{Phenol}}{C_6H_5OH} + \underset{\text{Acetone}}{(CH_3)_2CO}$$

Benzene Cumene Hydroperoxide of cumene Phenol Acetone

PROPERTIES OF ALDEHYDES AND KETONES

Aldehydes and ketones have similar chemical properties because they both contain a carbonyl group ($>C{=}O$), but generally, the aldehydes are more reactive than the ketones.

Order of Reactivity of Carbonyl Compounds

$$\underset{\substack{\text{formaldehyde}\\\text{(most reactive)}}}{\mathrm{H_2C{=}O}} > \underset{\text{acetaldehyde}}{\mathrm{CH_3(H)C{=}O}} > \underset{\text{acetone}}{\mathrm{(CH_3)_2C{=}O}} > \underset{\text{butan-2-one}}{\mathrm{C_2H_5(CH_3)C{=}O}}$$

$$\underset{\text{pentan-3-one}}{\mathrm{(C_2H_5)_2C{=}O}}\text{, etc. (least reactive)}$$

REACTIONS

(1) Aldehydes and ketones may be reduced to alcohols (this is the reverse of the reactions described above for the preparation of aldehydes and ketones)

e.g.
$$\underset{\text{acetaldehyde}}{CH_3CHO} + H_2 \xrightarrow[150^\circ C]{\text{Ni cat}} \underset{\text{ethanol}}{CH_3CH_2OH}$$

$$\underset{\text{acetone}}{CH_3CO{\cdot}CH_3} + H_2 \xrightarrow[150^\circ C]{\text{Ni cat.}} \underset{\text{propan-2-ol}}{CH_3CHOH{\cdot}CH_3}$$

(2) Using an oxidizing agent such as potassium dichromate in acid solution, aldehydes may easily be converted into the corresponding acids

e.g.
$$\underset{\text{acetaldehyde}}{CH_3CHO} + [O] \xrightarrow[\text{acid}]{K_2Cr_2O_7} \underset{\text{acetic acid}}{CH_3COOH}$$

Ketones cannot easily be oxidized. Strong oxidizing agents convert ketones to a mixture of acids, or ultimately to carbon dioxide and water.

Tollen's reagent and Fehling's solution (see the experiments at the end of this Chapter) are used to detect reducing agents, particularly aldehydes and substances containing aldehydic groups, e.g. glucose (p. 84)

e.g.
$$CH_3CHO + \underset{\text{Tollen's reagent}}{2Ag(NH_3)_2^+} + 2OH^- \rightarrow 2Ag\downarrow + CH_3COOH + 4NH_3 + H_2O$$

$$CH_3CHO + \underset{\text{Fehling's solution}}{2Cu(OH)_2} \rightarrow Cu_2O\downarrow + CH_3COOH + H_2O$$

(3) Aldehydes and ketones react with phosphorus halides to form compounds known as gem-dihalides (Latin: *geminus*, twin). In these substances both halogen atoms are attached to the same carbon atom.

$$CH_3CHO + PCl_5 \rightarrow CH_3\overset{\displaystyle H}{\underset{\displaystyle Cl}{\overset{|}{\underset{|}{C}}}}\!-\!Cl + POCl_3$$

acetaldehyde phosphorus pentachloride 1:1-dichloroethane

(4) Aldehydes and ketones undergo an important series of reactions known as *addition* and *condensation* reactions.

Addition Reactions

The carbonyl group reacts with certain reagents to yield only *one* product.

$$CH_3C\begin{matrix} \diagup\!\!\diagup O \\ \diagdown H \end{matrix} + NaHSO_3 \rightarrow CH_3\overset{\displaystyle H}{\underset{\displaystyle OH}{\overset{|}{\underset{|}{C}}}}\cdot SO_3Na$$

acetaldehyde sodium bisulphite acetaldehyde bisulphite

$$\underset{\text{acetone}}{CH_3CO\cdot CH_3} + NaHSO_3 \rightarrow CH_3\overset{\displaystyle OH}{\underset{\displaystyle SO_3Na}{\overset{|}{\underset{|}{C}}}}\cdot CH_3$$

acetone bisulphite

Acetaldehyde and acetone bisulphites are insoluble white solids. The reaction is important in that it may be used to separate carbonyl compounds from non-carbonyls. The carbonyl compound can be regenerated from the bisulphite by treating the latter with dilute acid.

Aldehydes and ketones will also add on hydrogen cyanide to form compounds called 'cyanhydrins'.

$$RCHO + HCN \rightarrow R{-}\underset{\underset{\textstyle CN}{|}}{\overset{\overset{\textstyle H}{|}}{C}}{-}OH$$

aldehyde cyanhydrin

$$R_2CO + HCN \rightarrow R{-}\underset{\underset{\textstyle CN}{|}}{\overset{\overset{\textstyle R}{|}}{C}}{-}OH$$

ketone cyanhydrin

In these reactions the reactants *add* together to form the product, hence the reason for calling them addition reactions.

Condensation Reactions

In these reactions of aldehydes and ketones *two* products are formed and one of these is always a simple substance such as water or ammonia

e.g. $\underset{\text{acetaldehyde}}{CH_3CHO} + \underset{\text{phenylhydrazine}}{H_2N{\cdot}NH{\cdot}C_6H_5} \rightarrow \underset{\text{acetaldehyde phenylhydrazone}}{CH_3CH{=}N{\cdot}NH{\cdot}C_6H_5} + H_2O$

$$\underset{\text{acetone}}{(CH_3)_2CO} + \underset{\text{phenylhydrazine}}{H_2N{\cdot}NHC_6H_5} \rightarrow \underset{\text{acetone phenylhydrazone}}{(CH_3)_2C{=}NNHC_6H_5} + H_2O$$

Reactions of this type are useful as a means of identifying particular aldehydes and ketones. The condensation products are often highly coloured and insoluble under the conditions of the reaction and may be filtered off and recrystallized from a suitable solvent. The pure material melts at a specific temperature which is characteristic of the condensation product and hence also of the original aldehyde or ketone.

USES OF ACETALDEHYDE, FORMALDEHYDE AND ACETONE

Acetaldehyde is used to make acetic acid, acetic anhydride, butanol, pentaerythritol [$C(CH_2OH)_4$—used in the manufacture of polyester resins] and crotonaldehyde ($CH_3CH{=}CHCHO$).

Most formaldehyde is used for making synthetic resins, e.g. Delrin, melamine and phenol-formaldehyde resins.

QUESTIONS

Acetone is used as a solvent for paints, varnishes, lacquers and cellulose acetate. It is also used to make other solvents, methyl methacrylate and drugs.

EXPERIMENTS

For the experiments described below use acetaldehyde (A) and acetone (B). Compare the reactivity of the two carbonyl compounds.

1. Add a few drops of Schiff's reagent to (A) and (B) separately. Notice in each case if there is any colour change.

2. Put 5 ml. of silver nitrate solution into a test-tube and add a drop or two of dilute sodium hydroxide solution. Add dilute ammonia solution until the grey precipitate *just* does not dissolve (this is known as Tollen's reagent and care must be taken not to allow it to evaporate dry since the residue is explosive). Add 1 ml. of (A). Notice the formation of a black precipitate of silver (sometimes a silver mirror is formed on the sides of the tube). Repeat using (B). Record what happens, if anything.

3. Add 5 ml. of Fehling's solution 1 (copper sulphate solution) to 5 ml. of Fehling's solution 2 (alkaline sodium potassium tartrate solution), then add 1 ml. of (A) and warm the mixture. Notice the formation of a brown precipitate. What is the brown precipitate? Explain the reaction which has occurred. Repeat using ketone (B). Record what happens.

4. Add 1 ml. of (A) to a few ml. of Brady's reagent (i.e. suspend $\frac{1}{2}$ g of 2:4-dinitrophenylhydrazine in 15 ml. of methanol and cautiously add a few drops of concentrated sulphuric acid to dissolve). Notice the formation of a red precipitate. Write the equation for the reaction. Repeat using (B). Describe what happens.

5. Prepare a strong solution of sodium bisulphite. To a few ml. of this solution add 2 ml. of (A). Repeat using (B). Describe what happens in each case and write the equations for the reactions.

QUESTIONS

1. What common property relates aldehydes and ketones?
2. Make a list of *all* the aldehydes containing up to six carbon atoms in their molecules. Give their names and write their structural formulae.
3. Name the following ketone:

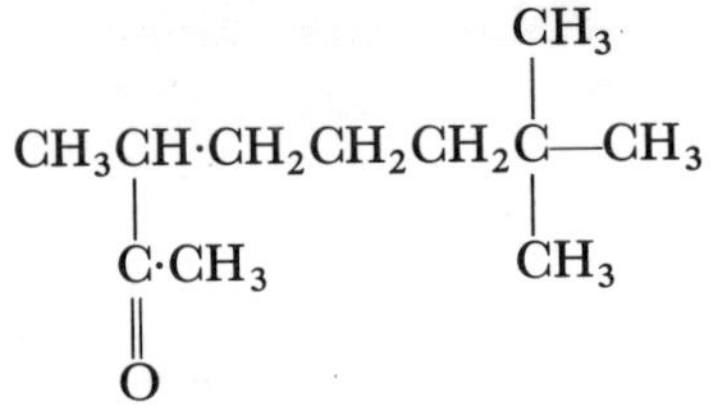

$$
\begin{array}{llr}
 & & CH_3 \\
 & & | \\
CH_3 & CH\cdot CH_2CH_2CH_2 & C\text{—}CH_3 \\
 & | & | \\
 & C\cdot CH_3 & CH_3 \\
 & \| & \\
 & O &
\end{array}
$$

4. Describe the laboratory preparation of a pure specimen of acetaldehyde.
5. How are acetaldehyde and acetone manufactured?
6. Describe what you would observe if (i) acetaldehyde, and (ii) acetone were treated with (a) Schiff's reagent, (b) Tollen's reagent, and (c) Fehling's solution.
7. Write equations for the reactions of (i) acetaldehyde (ii) acetone with phenylhydrazine. Name the products and explain why this reagent is important in organic chemistry.
8. How is it possible to separate benzaldehyde from ethanol using sodium bisulphite?
9. Explain the meaning of the terms addition reaction and condensation reaction.
10. Outline the commercial uses of acetaldehyde and acetone.

8

HALOGEN DERIVATIVES OF THE ALKANES

INTRODUCTION

THE simplest organic halogen compounds are the alkyl halides. These form a homologous series and may be represented by the general formula RX (where R = an alkyl group; X = Cl, Br, or I).

Table 8.1

Alkyl Halide	*Formula*
Methyl iodide	CH_3I
Ethyl iodide	C_2H_5I
1-Iodopropane (*n*-propyl iodide)	$CH_3CH_2CH_2I$
2-Iodopropane (*iso*propyl iodide)	CH_3CHICH_3
1-Iodobutane (*n*-butyl iodide)	$CH_3CH_2CH_2CH_2I$
2-Iodobutane	$CH_3CH_2CHI \cdot CH_3$
1-Iodo-2-methylpropane (*iso*butyl iodide)	$(CH_3)_2CHCH_2I$
2-Iodo-2-methylpropane	$CH_3C(CH_3)ICH_3$

The iodides are generally more reactive than the chlorides and bromides and the lower-member alkyl halides are gases or volatile liquids at room temperature. They are insoluble in water.

Table 8.2

Alkyl Halide	*Boiling point,* °C
Methyl chloride	−24
Methyl bromide	3
Methyl iodide	42

PREPARATION OF ALKYL HALIDES

Alkyl halides can be made from alcohols by replacing the hydroxyl group by a halogen atom.

$$\underset{\text{alcohol}}{ROH} \rightarrow \underset{\text{alkyl halide}}{RX}$$

(a) Alkyl chlorides may be prepared by passing dry hydrogen chloride into the alcohol in the presence of anhydrous zinc chloride (Grove's process)

e.g. $$\underset{\text{ethanol}}{C_2H_5OH} + HCl(\text{dry}) \xrightarrow[\text{anhydrous}]{ZnCl_2} \underset{\text{ethyl chloride}}{C_2H_5Cl} + H_2O$$

(b) Alkyl bromides can be made by refluxing (*Figure 8.1*) the alcohol with excess hydrobromic acid (48% in water)

e.g. $$\underset{\text{ethanol}}{C_2H_5OH} + HBr \xrightarrow[\text{conc.}]{H_2SO_4} \underset{\text{ethyl bromide}}{C_2H_5Br} + H_2O$$

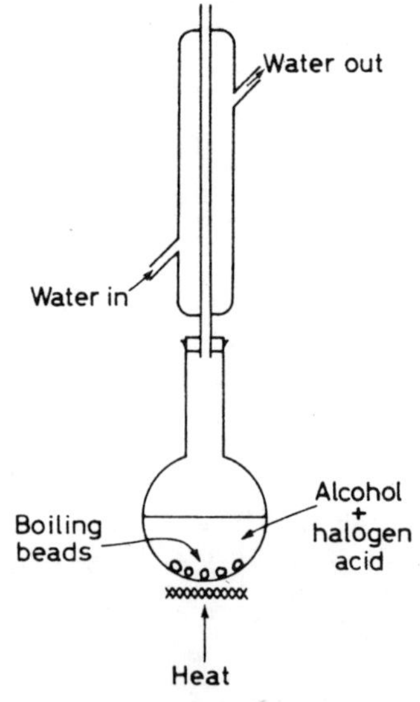

Figure 8.1

(c) Alkyl iodides can be prepared by refluxing the alcohol with excess hydroiodic acid (57% in water).

$$\underset{\text{propan-1-ol}}{C_3H_7OH} + HI \rightarrow \underset{\text{1-iodopropane}}{C_3H_7I} + H_2O$$

In all of these preparations the alkyl halide, having been made, must be separated from water and any unreacted alcohol or halogen acid. The usual procedure for separation involves:

(i) emptying the reaction mixture into a separating funnel and running off the alkyl halide (the reaction mixture forms two layers, one consisting of alkyl halide the other of alcohol plus halogen acid and water).

(ii) shaking the alkyl halide with sodium carbonate solution to destroy any excess acid and separating off the alkyl halide.

(iii) mixing the alkyl halide with calcium chloride solution to remove alcohol, and again separating off the alkyl halide.

(iv) drying the alkyl halide over solid anhydrous calcium chloride and finally distilling to obtain the pure product (*Figure 8.2*).

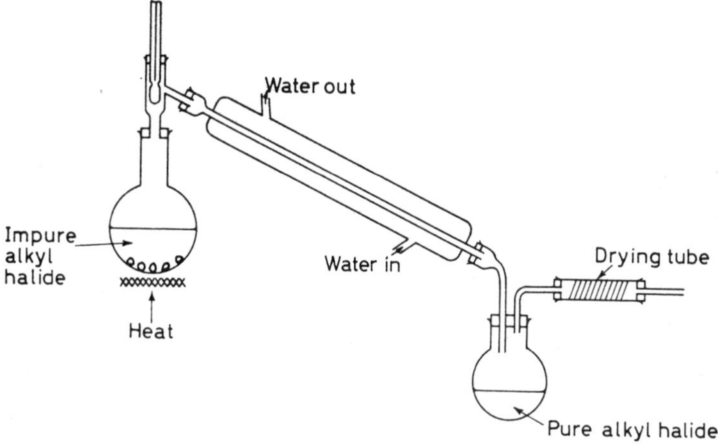

Figure 8.2

REACTIONS OF ALKYL HALIDES

(1) Alkyl halides can be reduced to alkanes. A number of reducing agents may be used, for example, a zinc–copper couple in alcohol, or sodium in alcohol

e.g. $$\underset{\text{ethyl bromide}}{C_2H_5Br} + 2[H] \rightarrow \underset{\text{ethane}}{C_2H_6} + HI$$

(2) Alkanes are also obtained by allowing alkyl halides to react with sodium in dry ether. This reaction is known by the name of the man who first investigated it, i.e. the Würtz reaction

e.g. $$\underset{\text{ethyl iodide}}{2C_2H_5I} + 2Na \rightarrow \underset{\text{butane}}{C_4H_{10}} + 2NaI$$

(3) When refluxed with aqueous alkali, the alkyl halides yield alcohols

e.g. $$\underset{\text{ethyl iodide}}{C_2H_5I} + KOH(aq.) \rightarrow \underset{\text{ethanol}}{C_2H_5OH} + KI$$

(4) Ethers (compounds containing the $\vdots C{-}O{-}C \vdots$ group) can be made from alkyl halides by treating the latter with alkoxides. The reaction is known as Williamson's synthesis

e.g. $$\underset{\text{ethyl iodide}}{C_2H_5I} + \underset{\text{sodium ethoxide}}{C_2H_5ONa} \rightarrow \underset{\text{diethyl ether}}{C_2H_5OC_2H_5} + NaI$$

(5) Alkyl halides can be converted to cyanides (also called nitriles) by reaction with potassium cyanide

e.g. $$C_2H_5I + KCN(alc.) \rightarrow \underset{\text{ethyl cyanide}}{C_2H_5CN} + NH_3$$

The cyanide is readily hydrolysed, by refluxing with dilute acid, to an organic acid.

$$C_2H_5CN + 2H_2O \rightarrow \underset{\text{propanoic acid}}{C_2H_5COOH} + NH_3$$

Note: These reactions afford a means of 'ascending' a homologous series since formation of a cyanide results in the introduction of a carbon atom into a molecule.

USES OF ALKYL HALIDES

Methyl chloride is used to make silicone polymers. These substances may be obtained in fluid, rubber or resin form and they are becoming of increasing importance as a result of their great chemical inertness. Fluid silicones have the added advantage of a very small coefficient of viscosity. This enables them to be used in very cold climates in place of ordinary lubricants.

Ethyl chloride is used for making tetraethyl lead. This substance is put into petrol to improve the combustion properties and prevent a phenomena caused by low-octane fuels known as 'knocking' (p. 23). Tetraethyl lead is made by two processes. One involves heating a lead–sodium alloy with ethyl chloride in an autoclave.

$$4C_2H_5Cl + 4Pb/Na \rightarrow (C_2H_5)_4Pb + 3Pb + 4NaCl$$

The other process is a modern one involving the electrolysis of an ethereal solution of ethyl magnesium chloride.

$$4[C_2H_5][MgCl] + Pb \xrightarrow{\text{electrolysis}} (C_2H_5)_4Pb + 2Mg + 2MgCl_2$$

Ethyl bromide is used as an intermediate in the manufacture of barbiturate drugs and 1-bromobutane is used as a stabilizer in the manufacture of polyvinyl chloride. Besides the uses mentioned above many alkyl halides are used extensively as solvents.

If the alkanes are further substituted by halogens, a number of very useful compounds are obtained, two of these being chloroform and carbon tetrachloride.

CHLOROFORM AND CARBON TETRACHLORIDE

Both chloroform and carbon tetrachloride are liquids and they are used extensively as solvents for many organic compounds.

Table 8.3

Substance	*Formula*	b.p., °C	m.p., °C
Chloroform	$CHCl_3$	61	−64
Carbon tetrachloride	CCl_4	77	−23

MANUFACTURE OF CHLOROFORM

There are two methods of manufacturing chloroform, one is by the action of bleaching powder on ethanol or acetone, the other is by reducing carbon tetrachloride.

(1) $$\underset{\text{ethanol}}{CH_3CH_2OH} + \underset{\text{from bleaching powder}}{Cl_2} \rightarrow \underset{\text{acetaldehyde}}{CH_3CHO} + 2HCl$$

$$CH_3CHO + 3Cl_2 \rightarrow \underset{\text{chloral}}{CCl_3CHO} + 3HCl$$

$$2CCl_3CHO + Ca(OH)_2 \rightarrow \underset{\text{chloroform}}{2CHCl_3} + \underset{\text{calcium formate}}{(HCOO)_2Ca}$$

(2) $$\underset{\text{carbon tetrachloride}}{CCl_4} + H_2 \rightarrow \underset{\text{chloroform}}{CHCl_3} + HCl$$

MANUFACTURE OF CARBON TETRACHLORIDE

Carbon tetrachloride is made industrially by chlorinating carbon disulphide using aluminium chloride catalyst.

$$CS_2 + 3Cl_2 \rightarrow CCl_4 + \underset{\text{sulphur monochloride}}{S_2Cl_2}$$

Carbon tetrachloride and sulphur monochloride are separated by fractional distillation.

REACTIONS

Chloroform and carbon tetrachloride are not particularly reactive. Of the two, chloroform is most reactive. As a result of its inertness carbon tetrachloride is used as a fire extinguisher (forms a blanket of vapour over the fire so preventing combustion).

(1) Chloroform is slowly oxidized in the presence of air and sunlight to phosgene (carbonyl chloride) which is a highly poisonous gas.

$$CHCl_3 + \tfrac{1}{2}O_2 \rightarrow \underset{\text{phosgene}}{COCl_2} + HCl$$

(2) Chloroform reacts with primary amines, when heated with strong alcoholic potash solution, to form compounds known as *iso*cyanides

e.g. $$CHCl_3 + 3KOH + C_6H_5NH_2 \rightarrow \underset{\text{phenyl } iso\text{cyanide}}{C_6H_5NC} + 3KCl + 3H_2O$$

This is an important reaction since it may be used to identify chloroform (or primary amines); *iso*cyanides have a very characteristic sickly smell.

(3) When boiled with strong aqueous or alcoholic solutions of sodium hydroxide, both chloroform and carbon tetrachloride are hydrolysed.

$$CHCl_3 + NaOH(aq.) \rightarrow \underset{\text{unstable}}{CH(OH)_3} \rightarrow \underset{\text{sodium formate}}{HCOONa}$$

$$CCl_4 + NaOH(aq.) \rightarrow \underset{\text{unstable}}{C(OH)_4} \rightarrow CO_2$$

USES OF CHLOROFORM AND CARBON TETRACHLORIDE

These substances are used extensively as solvents in a variety of industrial processes.

Chloroform is also used in the manufacture of polytetrafluoroethylene (P.T.F.E.).

$$CHCl_3 + 2HF \rightarrow CHClF_2 + 2HCl$$

$$2CHClF_2 \xrightarrow{700^\circ C} CF_2{=}CF_2 + 2HCl$$

$$n(CF_2{=}CF_2) \xrightarrow[\text{pressure}]{\text{catalyst}} (CF_2{-}CF_2)_n$$

The plastic, P.T.F.E., is important because of its chemical inertness, antistick properties and heat resistance. It shows great promise as a material for building chemical plant.

QUESTIONS

Carbon tetrachloride, besides being used as a solvent, is used to make refrigerants (dichlorodifluoromethane) and as a fire extinguisher.

EXPERIMENTS

Alkyl Halides

1. Shake ethyl bromide with:
(a) water
(b) cold dilute alkali
(c) cold dilute acid.
Say whether the alkyl halide is soluble in each case. Try and explain the result.

2. (a) To 1 ml. of ethyl bromide add 1 ml. of dilute nitric acid and 2 ml. of dilute silver nitrate solution. Shake. What happens? Warm the mixture. Is there any change?

(b) Add 2 ml. of dilute sodium hydroxide solution to 1 ml. of ethyl bromide in a test-tube. Boil the mixture. Allow to cool. Acidify with dilute nitric acid and add dilute silver nitrate solution. Notice whether a precipitate is formed. Explain the results of the experiments.

3. Dip a piece of copper wire into some ethyl bromide and heat the wire in a bunsen flame. Repeat a number of times. Record what you observe and explain the result.

Chloroform and Carbon Tetrachloride

1. Solubility: Test the solubility of the following substances in (i) carbon tetrachloride, and (ii) chloroform: Sodium chloride, naphthalene, water, glucose, benzoic acid and urea.

Can you explain why some of these substances are insoluble whilst others are soluble?

2. Put 1 ml. of ethanol and 0·5 ml. of chloroform in a test-tube, add two pellets of potassium hydroxide and warm gently shaking all the time. Acidify the mixture with dilute nitric acid and add silver nitrate solution. What is the white precipitate and what happens to the chloroform?

Repeat using carbon tetrachloride instead of chloroform.

QUESTIONS

1. How may ethyl chloride be prepared? Devise a suitable apparatus for the laboratory preparation.
2. Describe the laboratory preparation of a pure specimen of 1-iodopropane.
3. Write equations for the reactions of ethyl iodide with (i) nascent

hydrogen, (ii) sodium in ether, (iii) aqueous potassium hydroxide, (iv) sodium ethoxide, and (v) alcoholic potassium cyanide solution.

4. Write equations showing how ethyl iodide may be converted to propionic acid (propanoic acid).
5. Write the formulae and names of the compounds formed when excess ethyl bromide reacts with ammonia dissolved in alcohol.
6. Outline the uses of alkyl halides.
7. How may chloroform and carbon tetrachloride be prepared?
8. What are chloroform and carbon tetrachloride used for?
9. How do chloroform and carbon tetrachloride react with caustic soda solution?
10. Why is benzene soluble in carbon tetrachloride but insoluble in water?

9

ORGANIC ANALYSIS

In this chapter we shall investigate the ways of analysing and identifying organic substances. The approach is very similar to that adopted in inorganic analysis.

PURIFICATION

The first task is to ascertain whether the material is a mixture or a single substance. This is relatively straightforward since in many cases it is sufficient to determine its melting or boiling point. If the substance melts, or boils, over a wide range of temperature then it is probably a mixture. A pure substance melts and boils at definite temperatures, these being characteristic of the substance and a criterion of purity. (Methods of determining melting points and boiling points are described in the experiments at the end of Chapter 2.)

If a substance is found to be impure it can usually be purified by *recrystallization*, or in the case of a liquid, by *distillation*.

Recrystallization

Example—Purification of acetanilide contaminated with carbon.

The mixture is put into a boiling-tube containing distilled water. The tube is heated and the water allowed to boil gently for a few minutes. The *hot* solution is then filtered. On cooling, crystals of pure acetanilide separate from the solution. These are filtered off, washed with cold water and dried.

Separation and purification is achieved in this case by dissolving the organic substance, acetanilide, in hot water and filtering while hot. Carbon is insoluble in hot water and remains in the filter paper.

In many cases, separation is not as easily achieved. It may be necessary to experiment with various solvents to find one in which the pure substance is only sparingly soluble in cold, but completely soluble in hot solvent.

Distillation

Example—Separation and purification of ethyl benzoate dissolved in ether i.e. C_6H_5—C(=O)—OC_2H_5 in C_2H_5—O—C_2H_5

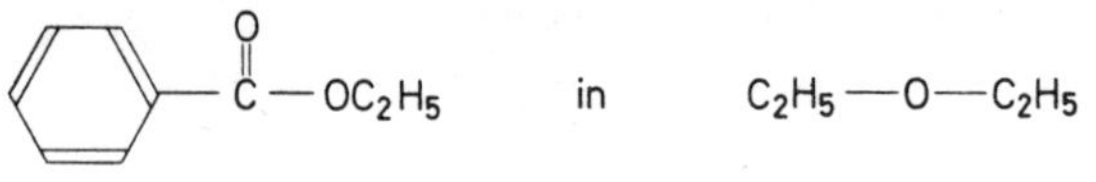

Ethyl benzoate — Diethyl ether (ether)

Ethyl benzoate and ether are liquids at room temperature. They are miscible with one another but since they have widely differing boiling points they may conveniently be separated by distillation (*Figure 2.2*). Ether boils at 35°C and ethyl benzoate at 209°C and therefore when the mixture is boiled ether distils off first. When all the ether has been removed the temperature is raised and the liquid distilling over in the range 207°–211°C is collected in a clean receiver. This is pure ethyl benzoate.

If the components of the liquid mixture have very similar boiling points then simple distillation is very often not sufficient to effect a separation. Under these circumstances fractional distillation is employed (p. 20).

Chromatography

Purification and Separation

In addition to the above methods, the substance may be investigated using chromatography. Chromatography means literally, colour writing and it was first used scientifically by the Russian botanist Tswett, in 1903, to separate the green colouring matter of plants, i.e. chlorophyll. In this case the components of chlorophyll were separated on a chalk column using petroleum ether as the elutant.

The principle behind this separation is that the components of chlorophyll have different solubilities in petroleum ether and show different attractions for the adsorbent. As a result the components travel through the column at varying speeds; the most soluble and least adsorbed travelling most rapidly.

By subjecting an unknown substance to a similar process as that described above for chlorophyll, it is possible to determine whether the substance is pure. If necessary the substance may be purified by separating and collecting the various components. If the components are not coloured, then other chromatographic methods may be employed.

To take an example, the naturally-occurring amino-acids may be separated chromatographically using paper as the adsorbent (*Figure 9.2*). Since the acids are colourless they cannot be detected until the paper is sprayed with locating agent. In this case the locating

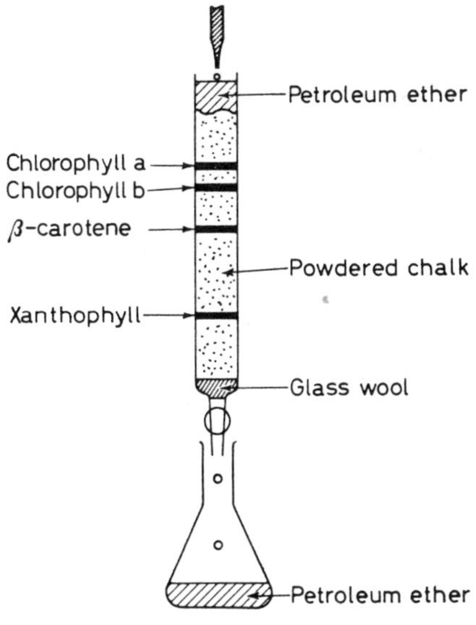

Figure 9.1

agent is a solution of ninhydrin in acetone. Ninhydrin reacts with the amino-acids to form purple-coloured complexes which are readily visible on paper.

Chromatography is most useful as a means of showing whether a substance is pure. It may also be used to effect a quantitative separation of the components of a mixture but this is very often a tedious business. Quantitative chromatographic separation is only employed when the components of a mixture are difficult to separate by simpler more conventional methods.

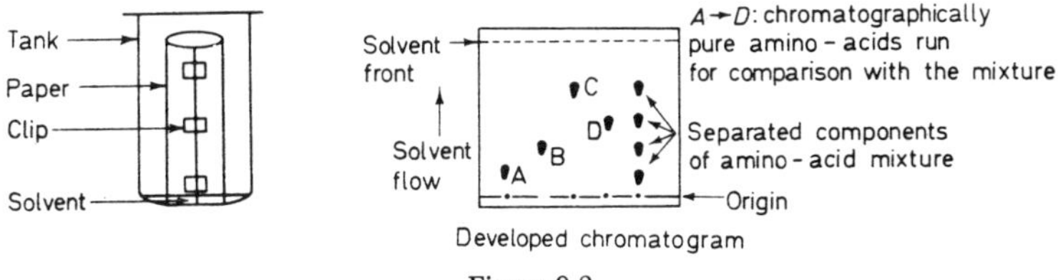

Figure 9.2

QUALITATIVE ANALYSIS

The next stage of the analysis, after purification, is concerned with identifying the elements which constitute the substance. This is referred to as qualitative analysis.

If the substance is organic then it must contain carbon and it is also most likely to contain hydrogen. The presence of these elements can be detected but they are usually assumed to be present and not tested for. In addition to carbon and hydrodgen, nitrogen, halogens, sulphur and oxygen are commonly found in organic compounds.

The elements nitrogen, halogens and sulphur may be tested for in the manner outlined below.

The substance is fused with sodium in an ignition tube. The tube is dropped, while hot, into a little distilled water contained in an evaporating basin. (*CAUTION:* When this is done, heating must be continued long enough to oxidize excess sodium before introducing the tube into distilled water. In addition, a wire gauze should be placed over the evaporating basin immediately the tube is introduced.) The water is boiled and filtered. The filtrate is divided into four portions and tested.

Table 9.1

Test	*Observation*	*Inference*
(1) A little ferrous sulphate solution is added to the filtrate. The mixture is boiled and acidified with dil. HCl	Blue colour or ppt.	Nitrogen present
(2) The filtrate is boiled for a few minutes with dilute nitric acid. The solution is diluted with distilled water and silver nitrate solution added	White ppt.	Halogen present (Cl, Br, I)
(3) A few ml. of carbon tetrachloride are added to the filtrate. Chlorine water is added and the mixture shaken. The carbon tetrachloride and aqueous layers are allowed to separate	Carbon tetrachloride layer brown	Bromine present
	Carbon tetrachloride layer violet	Iodine present
	No change indicates that original ppt. formed on addition of silver nitrate is due to silver chloride	Chloride present
(4) 2% aqueous solution of sodium nitroprusside added to filtrate	Deep violet colour	Sulphur present

A knowledge of the elements present in an organic compound enables us to make certain deductions concerning its chemical nature. For example, if nitrogen is found to be present, the substance may be an amine or amide. If it contains halogens as well as nitrogen, it may be an amine salt. If it is found that nitrogen, halogens and sulphur are *absent* then the substance may belong to one of the following groups; acids, acid anhydrides, aldehydes, ketones, alcohols, esters, or carbohydrates. In order to discover the functional groups present in a molecule of the compound, some of the simple tests and reactions that have been considered in the previous chapters must be applied.

QUANTITATIVE ANALYSIS AND IDENTIFICATION

Following the approach outlined above, the analyst is able to describe the chemical properties of the substance fairly accurately. He will be able to say whether it is complex or simple and what functional groups it contains. He is not able, at this stage, to identify the substance positively. To achieve this the substance must be analysed quantitatively and its molecular weight determined. Finally the substance must be synthesized and the product shown to be identical with the original unknown substance.

The final stages of analysis are best illustrated by considering specific examples.

Example 1

A pure organic compound was found to contain carbon, hydrogen and bromine only. Quantitative analysis revealed that the substance contained these elements in the proportions, C = 12·77%, H = 2·13% and Br = 85·1%. Its molecular weight by Victor Meyer's method was found to be 188.

On hydrolysis with aqueous potassium hydroxide it yielded ethylene glycol. Quantitative analysis revealed that the substance is relatively simple and only contains carbon, hydrogen and bromine.

In every 100 g of the substance there was 12·77 g of carbon, 2·13 g of hydrogen and 85·1 g of bromine.

∴ the atomic ratio of these elements is:

	C	:	H	:	Br
	12·77/12		2·13/1		85·1/80
	1·064		2·13		1·064
or	1		2		1

i.e. CH_2Br

This is known as the *empirical formula*. It only indicates the type and ratio of atoms comprising the substance.

The *molecular formula* must now be worked out. This indicates how many atoms of each of the elements are present in one molecule of the substance.

The empirical formula is CH_2Br. If this was also the molecular formula the substance would have a molecular weight of;

$$12 + (2 \times 1) + 80 = 94$$

The molecular weight is in fact 188, i.e. 94×2

$$\begin{aligned} \therefore \text{ molecular formula} &= 2(CH_2Br) \\ &= \underline{C_2H_4Br_2}. \end{aligned}$$

This may correspond to the structures;

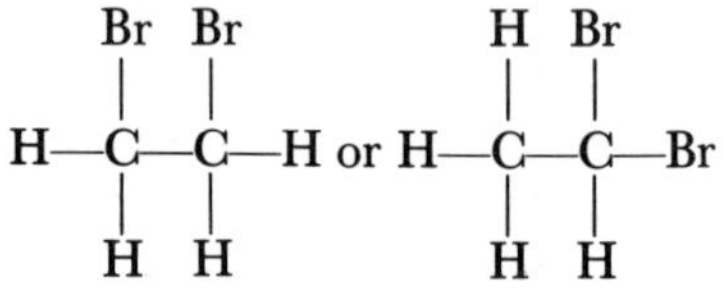

$$\begin{array}{ccccc} & Br & Br & & \\ & | & | & & \\ H- & C & -C- & H & \end{array} \text{ or } \begin{array}{ccccc} & H & Br & & \\ & | & | & & \\ H- & C & -C- & Br & \end{array}$$

$$\begin{array}{cc} H & H \end{array} \qquad\qquad \begin{array}{cc} H & H \end{array}$$

1:2-dibromoethane 1:1-dibromoethane

(I) (II)

On hydrolysis I yields 1:2-dihydroxyethane.

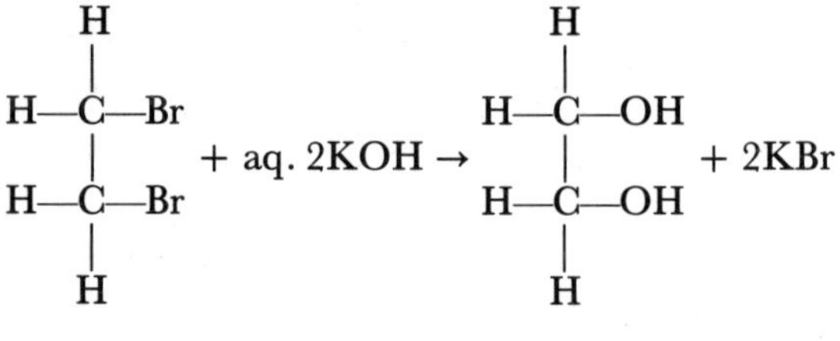

$$\begin{array}{c} H \\ | \\ H-C-Br \\ | \\ H-C-Br \\ | \\ H \end{array} + \text{aq. } 2KOH \rightarrow \begin{array}{c} H \\ | \\ H-C-OH \\ | \\ H-C-OH \\ | \\ H \end{array} + 2KBr$$

Substance II would give acetaldehyde on hydrolysis.

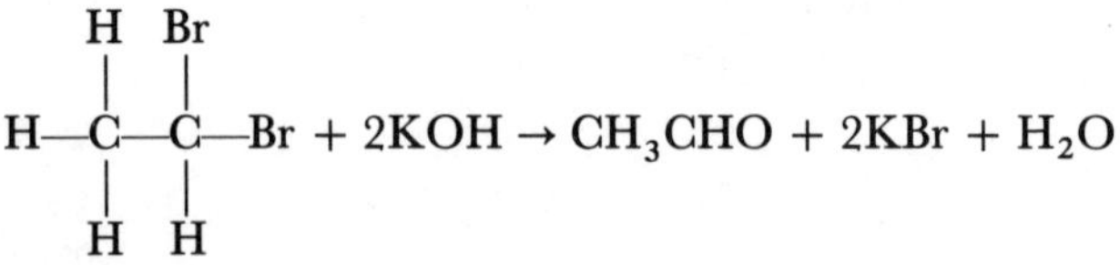

$$\begin{array}{cccc} & H & Br & \\ & | & | & \\ H- & C & -C- & Br \\ & | & | & \\ & H & H & \end{array} + 2KOH \rightarrow CH_3CHO + 2KBr + H_2O$$

It follows therefore that the unknown substance is 1:2-dibromoethane.

To prove conclusively that the substance is 1:2-dibromoethane it may be compared with an authentic sample of the substance synthesized in the following way.

$$\underset{\text{ethene}}{\begin{array}{c}CH_2\\ \| \\ CH_2\end{array}} + Br_2 \rightarrow \underset{\text{1:2-dibromoethane}}{\begin{array}{c}CH_2Br\\ | \\ CH_2Br\end{array}}$$

Example 2. The Structure of Cocaine

Pure cocaine was first used as a pain reliever in 1884. Since this time its structure has been elucidated with a view to synthesizing other substances having the same anaesthetic properties but none of the undesirable properties of cocaine (p. 116).

The methods adopted for its analysis were as follows. The pure substance was partially degraded (i.e. broken down into smaller molecules) by mild hydrolysis to methanol and benzoyl ecgonine and then by more powerful hydrolysis to ecgonine, benzoic acid and methanol.

$$\underset{\text{cocaine}}{C_{17}H_{21}NO_4} + H_2O \xrightarrow[\text{hydrolysis}]{\text{mild}} \underset{\text{benzoyl-ecgonine}}{C_{16}H_{19}NO_4} + \underset{\text{methanol}}{CH_3OH}$$

$$\underset{\text{cocaine}}{C_{17}H_{21}NO_4} + 2H_2O \xrightarrow[\text{hydrolysis}]{\text{powerful}} \underset{\text{ecgonine}}{C_9H_{15}NO_3} + \underset{\text{benzoic acid}}{C_7H_6O_2} + CH_3OH$$

From the results of these experiments it was deduced that cocaine is the methyl ester of benzoyl ecgonine.

The chemical properties of ecgonine indicate that it contains a secondary alcohol, an acid and a tertiary amine group.

$$\underset{\text{secondary alcohol group}}{>CHOH}, \quad \underset{\text{acid group}}{-C{\overset{O}{\underset{OH}{\lessgtr}}}}, \quad \underset{\text{tertiary amine group}}{>N-}$$

Reaction with zinc dust yields 2-ethylpyridine and careful oxidation with potassium permanganate yields *N*-methylsuccinimide.

$$\text{Ecgonine} \xrightarrow{Zn} \text{2-ethylpyridine (pyridine ring, positions 1–6, } C_2H_5 \text{ at position 2)}$$

$$\text{Ecgonine} \xrightarrow{[O]} \begin{array}{l} H_2C-C(=O) \\ \;\;|\qquad\qquad >N-CH_3 \\ H_2C-C(=O) \end{array} \quad N\text{-methylsuccinimide}$$

These reactions indicate that ecgonine has the structure

```
       CH———————CH·COOH
H2C               
 |       N—CH3   CHOH
H2C                       Ecgonine
       CH———————CH2
```

By piecing together the results of these analyses, cocaine can be shown to have the structure

```
       CH———————CH·COOCH3
H2C
 |       N—CH3   CH·O·CO C6H5
H2C
       CH———————CH2
```

Final justification for this was obtained by the synthesis of a substance having this structure which was then shown to be identical to natural cocaine.

Today much of the hard labour of chemical analysis has been eliminated by the development of various types of instrumental analysis. Of these, probably the most powerful are infra-red, ultraviolet, nuclear magnetic resonance and mass spectroscopy. These devices in the hands of experienced technicians are able to give information about the structure of substances which would otherwise take weeks or months to discover.

EXPERIMENTS

1. Ask the chemistry teacher to give you an organic substance to analyse for nitrogen, halogens and sulphur (refer to the text for details of the analysis. *Note*: Use a piece of sodium about half the size of a dried pea and take care to ensure that excess sodium is oxidized (by heating in air) before plunging the tube into distilled water).

2. *Chromatography*: Set up the apparatus shown in *Figure 9.3*. Spot a concentrated ink solution onto the paper and allow the paper to absorb the solvent (use a fine capillary tube to spot the mixture). The ink solution may be variously-coloured fountain-pen or biro inks dissolved in a little methanol. Use methanol as solvent in the first place then try other alcohols and alcohol–water mixtures. Discover the number of coloured components comprising the inks you use.

Repeat the experiments using a chlorophyll extract. Crush some

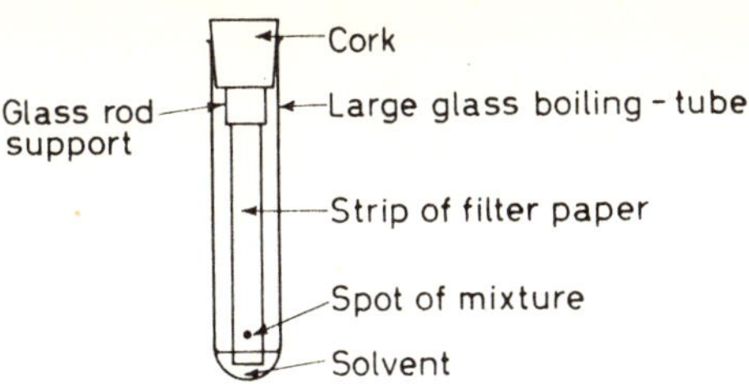

Figure 9.3

green leaves in a mortar with some acetone. Dry the extract by adding a few grammes of anhydrous sodium sulphate and allowing the mixture to stand for a few minutes, then filter. Spot the filtrate onto the filter paper (see *Figure 9.3*). Soak the paper with acetone–petroleum ether (40–60) mixed in the proportion 10:90. (Take care to keep this solvent away from naked flames, it is *very* inflammable).

3. *Recrystallization*. Prepare the following solutions:

(a) 1 g of sodium acetate in 1 ml. of water.
(b) $\frac{1}{2}$ g of semicarbazide hydrochloride in 2 ml. of water.
(c) 1 ml. of benzaldehyde in 3 ml. of alcohol.

Mix solutions (a) and (b) then add to (c). A white precipitate of benzaldehyde semicarbazone is formed.

$$\underset{\text{benzaldehyde}}{C_6H_5CHO} + \underset{\substack{\text{semicarbazide}\\\text{hydrochloride}}}{Cl^-NH_3^+NH\ CONH_2} \xrightarrow[\text{acetate}]{\text{sodium}}$$

$$\underset{\text{benzaldehyde semicarbazone}}{C_6H_5CH{=}N.NHCONH_2\downarrow} + HCl + H_2O$$

Filter off the precipitate and put it into a clean test-tube. Just cover the precipitate with 50:50 water–ethanol mixture and warm the tube in a beaker of hot water. If the semicarbazone does not dissolve after a few minutes add a little more aqueous alcohol and warm the solution again. Repeat this procedure until the solid *just* dissolves then cool the solution. Crystals of the semicarbazone will be seen to appear as the solution attains room temperature. Filter off the crystals and dry them in the oven (at about 100°C). When dry, determine the melting point and compare it with the value given in a reference book of physical constants.

QUESTIONS

1. Outline briefly how you would attempt to analyse and identify an organic substance.

2. A pure organic compound, containing carbon and hydrogen only, was analysed quantitatively. The following results were obtained.

$$\text{carbon} = 87\%$$
$$\text{hydrogen} = 13\%$$

Calculate the empirical formula of the substance.

3. A pure organic compound was analysed quantitatively and found to contain 21·17% carbon, 4·12% hydrogen and 74·71% iodine. Its molecular weight was found to be 170. Calculate the molecular formula and write equations for the reactions of the isomers with aqueous potassium hydroxide solution.

4. On analysis a pure organic liquid was found to contain 37·5% carbon, 12·5% hydrogen and 50% oxygen. Calculate its empirical formula and assuming this to be the molecular formula, write the structural formula.

5. A pure white crystalline organic substance was analysed quantitatively. Carbon, hydrogen and nitrogen were found to be present in the following proportions.

$$\text{carbon} = 20{\cdot}00\%$$
$$\text{nitrogen} = 46{\cdot}66\%$$
$$\text{hydrogen} = 6{\cdot}66\%$$

The molecular weight of the substance was found to be 60. Calculate its molecular formula and write possible structural formulae.

PART III

10

FOODS

OUR food consists of five important types of compounds namely, carbohydrates, fats, proteins, vitamins and salts. All of these substances are essential if the body is to function properly. In the following pages the nature and action of these essential constituents of our diet will be discussed.

CARBOHYDRATES

When our bodies are working hard and regularly, the carbohydrate component of the food we eat is digested in the stomach and intestines, and converted to simple sugars (e.g. glucose). These simple sugars pass from the alimentary canal into the main blood system (via capillaries) and are conveyed to the muscles and other parts of the body, where they function as fuel. If food is taken in and the muscles are not required to do work, then excess sugar is converted to a substance called glycogen ('animal starch'). This is mostly stored in the liver and can be converted back into simple sugars when required by the body. Animals, unlike some plants (e.g. potatoes, oak and horse-chestnut) are not able to store excessive amounts of polysaccharide (starch) as reserve fuel, but they are able to store fat which can also be converted into simple sugars when required.

Carbohydrates are organic substances containing carbon, hydrogen and oxygen; the hydrogen and oxygen is present in the proportion 2:1 (the same as in water) and carbohydrates may be represented by the general formula, $C_x(H_2O)_y$. Sucrose is a carbohydrate and has the general formula, $C_{12}H_{22}O_{11}$.

The simplest carbohydrates are called *monosaccharides*, glucose and fructose being examples.

CHO	CH_2OH
|	|
CHOH	CO
|	|
$(CHOH)_3$	$(CHOH)_3$
|	|
CH_2OH	CH_2OH
glucose	fructose

Sucrose, maltose and lactose are examples of more complex carbohydrates called *disaccharides* since they consist of two monosaccharides joined together.

Sucrose has the general formula:

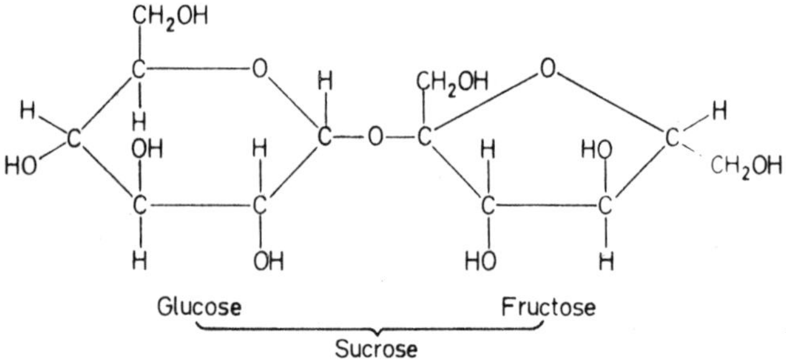

Glucose Fructose

Sucrose

Starch and cellulose are examples of still more complex carbohydrates. They have a very high molecular weight (10,000–500,000) their molecules consisting of numerous monosaccharide units joined together.

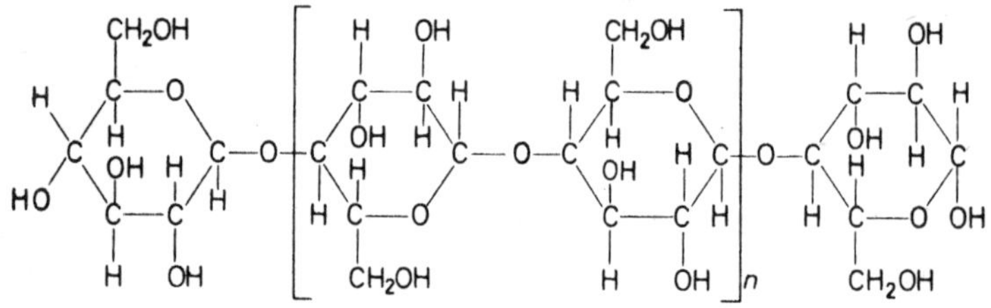

($n \simeq 1000$)

Cellulose

Cellulose constitutes the fibrous parts of plants and although it cannot be digested in our bodies, it may be digested by animals, which feed on grass, and in their bodies converted to sugars.

FATS

Fats consist mostly of complex esters (p. 45) and they are utilized by the body as a source of chemicals for building new tissue or to produce energy for vital processes. In the alimentary canal fats are converted, by the action of enzymes, to fatty acids and glycerol.

$$\begin{array}{lcl} CH_2OOCR & & CH_2OH \\ | & & | \\ CHOOCR + 3H_2O & \xrightarrow{\text{enzymes}} & CHOH + 3RCOOH \\ | & & | \qquad\qquad\quad \text{fatty acid.} \\ CH_2OOCR & & CH_2OH \quad \begin{array}{l} R = \text{alkyl group,} \\ \text{e.g., } C_{17}H_{35}\text{-stearyl} \end{array} \\ \text{fat} & & \text{glycerol} \end{array}$$

Glycerol and fatty acids then pass out of the intestines and into the blood via the lymph system. In the lymphatic vessels glycerol and fatty acids are combined to form minute globules of fat. The fat is then conveyed to the blood (by way of the thoracic duct) and transported to the various tissues where it provides a source of energy, or is used to build up new protoplasm.

PROTEINS

Proteins are complex substances of high molecular weight which contain carbon, hydrogen, oxygen and nitrogen. We may classify them into three broad groups.

(1) Proteins used by the body as building material because of their fibrous nature.

(2) Proteins having specific uses as enzymes, hormones, and nucleoproteins.

(3) Food proteins from which the essential amino-acids are derived. (These amino acids are used in their turn to synthesize proteins.)

Proteins belonging to the first two groups are made by the body from proteins contained in food (group 3). In the process of digestion the protein in food is broken down by the action of enzymes, in the stomach and intestines (i.e. pepsin in the stomach, and trypsin in the intestines), to amino-acids. The amino-acids are then reformed into the proteins required by the body.

Amino-acids are relatively simple organic compounds. They have the general formula:

$$\begin{array}{c} H \\ | \\ R\text{—}C\text{—}C \begin{array}{l} {}^{\diagup\!\!\diagup}O \\ {}_{\diagdown}H \end{array} \\ | \\ NH_2 \end{array} \qquad \begin{array}{ll} \text{—}NH_2 & \text{Amino-group} \\ \text{—}COOH & \text{Carboxyl group} \end{array}$$

Amino-acid

The group R can vary considerably and, in nature, there are approximately 25 different R groups and hence 25 different amino-acids. For example:

Amino-acid	R Group
Threonine	$CH_3CHOH—$
Phenylalanine	$C_6H_5CH_2—$
Lysine	$NH_2—(CH_2)_3—CH_2—$
Tryptophan	(indole ring, N–H) $CH_2—$
Valine	$(CH_3)_2—CH—$
Cysteine	$HS—CH_2—$
Leucine	$(CH_3)_2—CH—CH_2—$
*iso*Leucine	$CH_3—CH_2—CH(CH_3)—$

In a protein molecule amino-acids are joined together through bonds between the carboxyl group of one amino-acid and the amino-group of another by the removal of water. The bond so formed is called a *peptide link*. When two amino-acids are joined in this way the resulting compound is called a *dipeptide*. Similarly a *tripeptide* contains three amino-acids joined together by peptide bonds and a *polypeptide* contains a large number of amino-acids joined together. For example:

$$\underset{\text{glycine}}{NH_2{\cdot}CH_2CO}\boxed{OH + H}\underset{\text{glycine}}{NHCH_2COOH} \rightarrow$$

$$\underset{\text{dipeptide}}{NH_2CH_2CONHCH_2CO}\boxed{OH + H}\underset{\text{glycine}}{NHCH_2COOH} \rightarrow$$

$$\underset{\text{tripeptide}}{NH_2CH_2CONHCH_2CONHCH_2COOH} + \text{etc.}$$

Protein molecules consist of peptides and polypeptides folded into unusual shapes and in many cases the structure is stabilized by cross-linking of adjacent peptides. These cross-links, holding peptide chains together, are very often disulphide bonds (—S—S—). Cysteine is an amino-acid which contains sulphur and the acid is a key member of these cross linkages. For example:

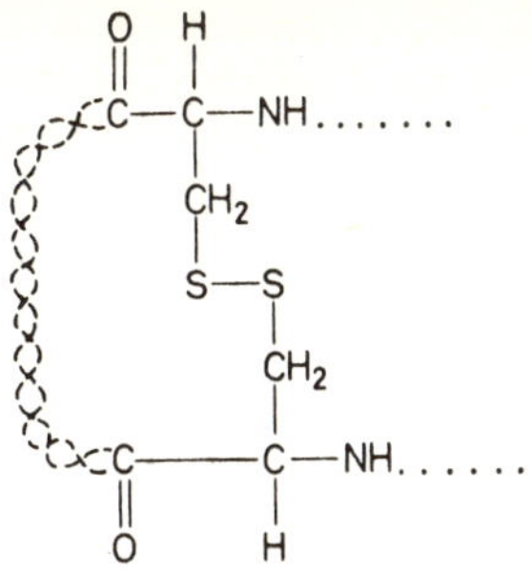

```
      O  H
      ‖  |
  ...C—C—NH.......
  :      |
  :      CH2
  :      |
  :      S—S
  :        |
  :        CH2
  :        |
  ...C—————C—NH......
     ‖     |
     O     H
```

VITAMINS

The diseases, rickets, beri-beri, pellagra, scurvy and some forms of dermatitis have one thing in common and that is, that they are non-infectious and are caused by a deficiency of certain substances in our diet. These substances are collectively known as *vitamins* so called because, when their significance was realized, they were thought to be vital amines. Vitamins, in fact, contain the functional groups of a variety of classes of compounds.

Today, we are aware of the importance of vitamins as a result of advertising and not so much by the physical disabilities resulting from their absence in our diet. In the time of Elizabeth I things were much different and thousands of British seamen died every year from scurvy, a disease caused by a deficiency of vitamin C in the diet. Vitamin C is one of the more essential vitamins and, although we only need about 30 mg per day, if our daily intake falls below about 10 mg for any length of time the symptoms of scurvy become apparent, i.e. wounds heal only slowly, teeth decay and skin softens and bleeds easily. Vitamin C is contained in appreciable quantity in fruit and vegetables.

Besides vitamin C there are four other vitamins, or groups of vitamins, which are important and these, like vitamin C, are denoted by capital letters, i.e. A, B, D and E. For convenience we may divide the vitamins into two groups: (1) those soluble in water, and (2) those soluble in fats. Those belonging to group (1) include vitamins B and C (also P). Those in group (2) include vitamins A, D and E (also K).

Group 1

Vitamin B

This vitamin is complex since it contains at least 12 components (B_1–B_{12}), each component serving a special purpose. Vitamin

B_1 (aneurin) deficiency results in nervous disorders and an inability to utilize carbohydrates efficiently. The vitamin was synthesized in 1936 when its structure was shown to be:

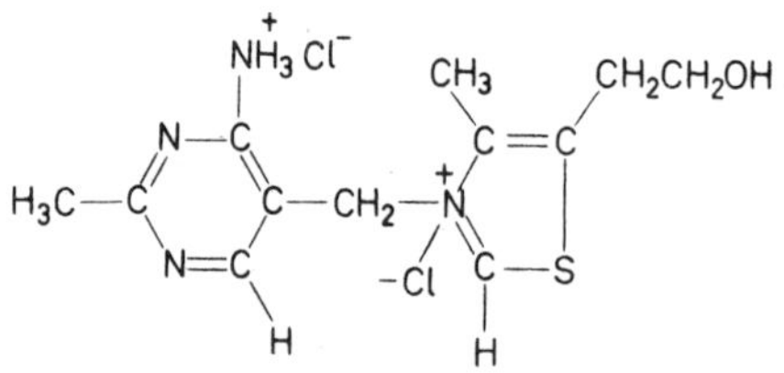

Vitamin B_1

Vitamin B_2 (riboflavin) has a yellow colour and is to be found in milk, liver and yeast. Its deficiency is the cause of the disease pellagra in which the skin and intestines become inflamed. The vitamin was discovered as a result of research into the cause of pellagra, and it was first synthesized in 1934. It has the structural formula:

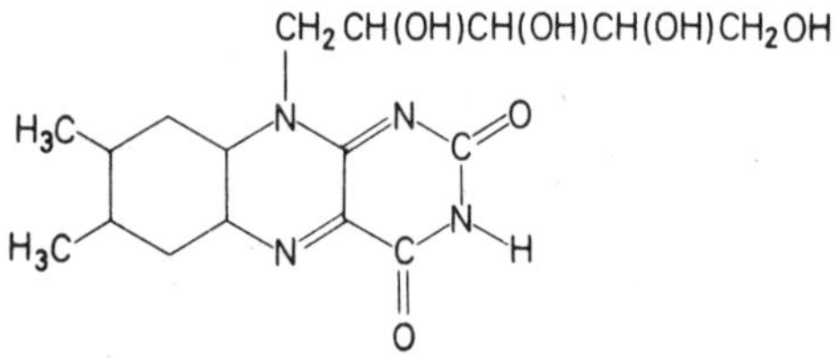

Vitamin B_2

To cure the disease pellagra, two vitamins must be administered together, i.e. vitamin B_2 and vitamin B_4 (nicotinic acid). Nicotinic acid is a relatively simple substance and has the structure:

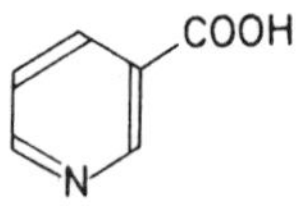

Vitamin B_4

It occurs in green vegetables, meat and yeast.

Vitamin C

This vitamin has been mentioned previously. It was synthesized in 1933 but was first used in the form of lemon juice as long ago as 1590 to cure a ship's company of scurvy. Vitamin C (ascorbic acid) has the structure:

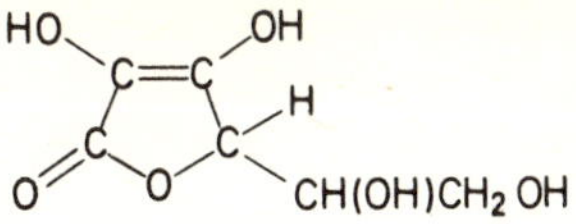

HO–C=C–OH, lactone ring: O=C–O–C(H)–CH(OH)CH$_2$OH

Occurring with vitamin C, and probably associated with it in some way, is vitamin P (hesperidin or citrin). Its structure is unknown.

Group 2

Vitamin A

Vitamin A is another of the important vitamins. A deficiency of this results in stunted growth in the young, night blindness and scaling of the skin. It occurs in cod and halibut oils, milk, eggs, butter, tomatoes and some vegetables such as carrots. The physiological behaviour of vitamin A (contained in milk) was discovered by Gowland Hopkins who, in 1912, observed that rats fed only on pure proteins, carbohydrates, fats, water and salts, lost weight but recovered this when fed with small doses of milk. It was not until 1937 that vitamin A was synthesized and its structure shown to be:

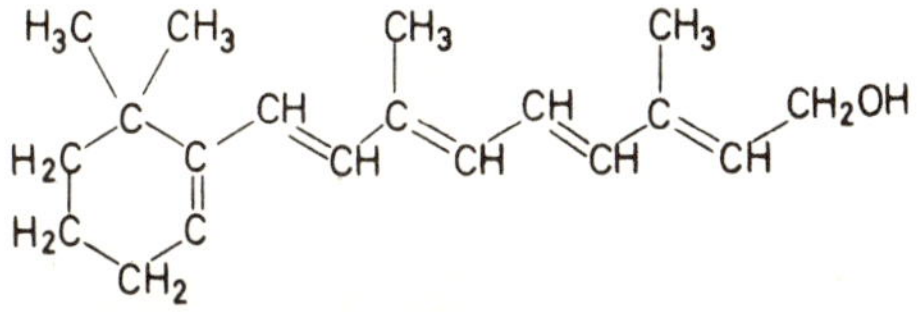

Vitamin A

Vitamin A may be obtained from a substance called β-carotene. This is orange-red in colour and is contained in carrots, egg-yolks, grass (see *Figure 9.1*) and certain marine creatures. Vitamin A is stored in the liver and in this organ β-carotene is converted to vitamin A.

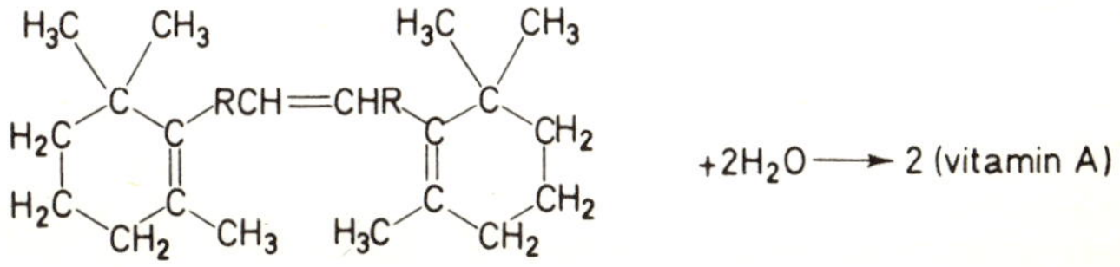

RCH=CHR $+2H_2O \longrightarrow$ 2 (vitamin A)

β–Carotene

Vitamin D

Vitamin D is concerned with the absorption of calcium and phosphorus from the intestines. These substances are used to make bone and maintain the skeleton in good repair. Before 1940 many children suffered from a disease known as rickets which is characterized by bowlegs and deformed spines and which is caused by lack of vitamin D.

Vitamin D has posed somewhat of a problem to chemists and biochemists since rickets, etc., caused by lack of the vitamin, can be cured in two ways: (1) irradiating the patient with sunlight or ultraviolet light (in small doses over a long period) or (2) administering cod or halibut oil. The problem was to discover the connection between these two remedies and find the actual chemical producing the cure. After a good deal of research it was found that a number of chemically very similar substances cure rickets. These are collectively known as vitamin D and will cure and prevent rickets if included in the diet. The action of sunlight is more complex; our skin contains substances called sterols, more specifically sterols containing a hydroxyl group in the 3 position (provitamins, see structure below), and it is these compounds which are converted to vitamin D by the action of sunlight.

We may summarize what has been said above as follows:

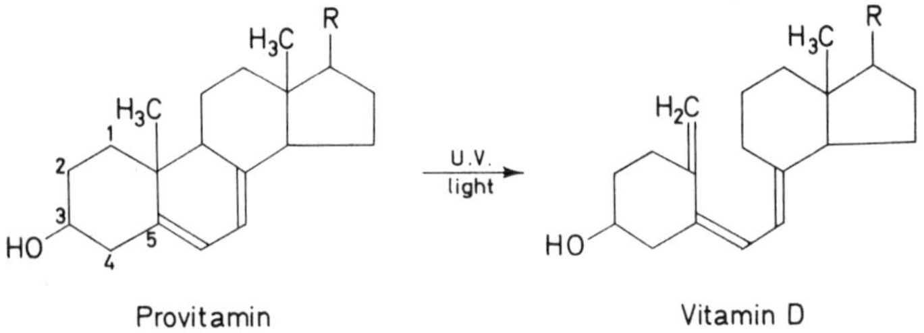

Provitamin → (U.V. light) → Vitamin D

R = long chain alkyl group

Vitamin E

Vitamin E is another complex vitamin. There are at least four chemicals which comprise vitamin E, i.e. α, β, γ and δ tocopherols (Greek: *tokos*, childbirth; *phero*, to bear). As the name tocopherol suggests the action of vitamin E is concerned with fertility in animals. It has been shown, for example, that in rats vitamin E deficiency causes sterility and muscular dystrophy.

The tocopherols are derivatives of 6-hydroxychroman (see below). α-Tocopherol has three methyl groups in the benzene nucleus, β and γ-tocopherols two methyl groups, and δ-tocopherol one methyl group in the benzene nucleus. The β and γ compounds differ in that the two methyl groups are differently situated in the benzene nucleus (β-methyls *para* to one another; γ-methyls on the same side of the —OH group).

Vitamin E is contained in wheat germ and green vegetables.

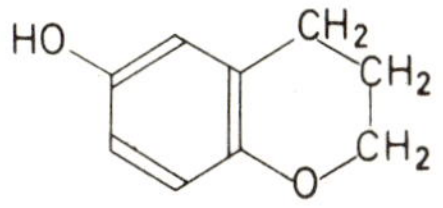

6-hydroxychroman

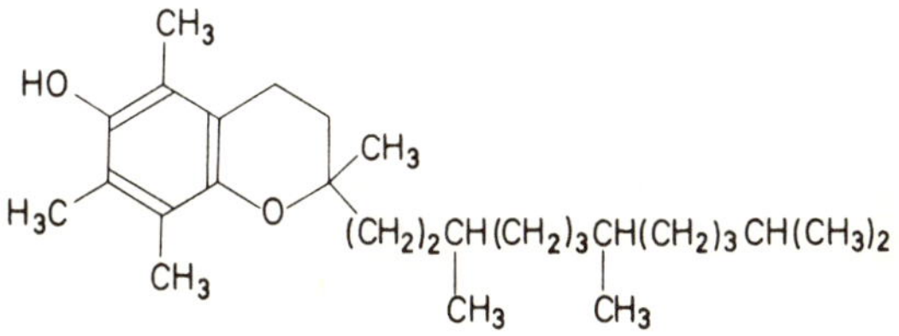

α-tocopherol

SALTS

Besides carbon-containing substances such as carbohydrates, fats and proteins, our food also contains inorganic salts. These salts provide elements, other than carbon, hydrogen, oxygen, nitrogen and sulphur, essential for good health. Calcium and phosphorus, for example, are required for building bones and teeth. (*Note*: Vitamin D is also needed when calcium and phosphorus are utilized for this purpose.) Calcium is also used to promote the action of some enzymes and phosphorus in building amino-acid-containing substances.

Every year our bodies use about eight kilogrammes of potassium and sodium salts. We are aware of these substances in our body fluids since when we sweat our skin tastes salty and tears also have a salty taste. It seems that salts serve a number of functions, the chemistry of which is a little obscure, but it is known for example that they are necessary for the proper functioning of the muscles.

Iron is contained in haemoglobin, which is the red substance in blood and which is responsible for carrying oxygen used in respiration. Although the ratio of iron atoms to carbon atoms in one molecule of haemoglobin is only 1:758, the substance will not carry oxygen if iron is not present. A deficiency of iron results in anaemia and general weakness. Good sources of iron are bread, potatoes and vegetables.

Another element which we might be surprised to find in the body is cobalt, which is contained in vitamin B_{12}. Vitamin B_{12} has the formula, $C_{63}H_{88}N_{14}O_{14}PCo$ and as one might expect it has a very complicated structure. The structure was finally elucidated in 1956 by Dr. Dorothy Hodgkins and her colleagues at Oxford University using a specialized technique known as X-ray analysis. Liver is a good source of cobalt.

Besides the elements mentioned above, the body is also known to utilize ten others, most of them present in extremely small amounts.

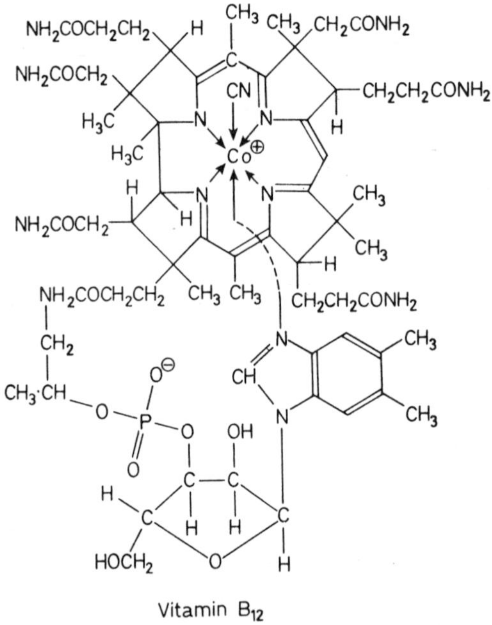

Vitamin B_{12}

11

POLYMERS

INTRODUCTION

The term polymer is used to denote a substance composed of very large molecules built from numerous identical chemical units. These large molecules are sometimes referred to as macromolecules, this being a general term applied to any large molecule containing repeating chemical units.

Protein, cellulose and rubber (Chapter 12) are examples of natural polymers. They are high-molecular-weight substances, their molecules, in many cases, being comprised of thousands of simple chemical units. Proteins, for example, are built from the relatively-simple amino-acids (Chapter 10) and cellulose from simple sugars (Chapter 10).

Various forms of cellulose such as cotton and hemp have been used for thousands of years for making fabrics. In an attempt to produce a synthetic material having similar physical properties to natural cellulose fibre, man has produced a variety of polymeric substances. Some of these have been found useful for making synthetic fibre to supplement or replace natural fibre but the majority are unsuitable for this purpose and have found other applications. Of these synthetic polymers, the better known include polyethylene (polythene), polyvinyl chloride (P.V.C.), polystyrene, polyurethane, Nylon and Terylene. These are described as 'thermo-softening' plastics because they soften when heated. On the other hand, Bakelite and the amino-plastics are referred to as 'thermo-setting' plastics because they cannot be remoulded by the application of heat.

PREPARATION AND USES OF SYNTHETIC POLYMERS

Thermo-softening Plastics

Polyethene (Polythene)

Polythene is made from ethene. When the gas is passed over a catalyst (e.g. chromium oxide) at 200°C and at low pressure the ethene molecules 'polymerize' (join together) to yield polythene.

$$CH_2{=}CH_2 + CH_2{=}CH_2 + CH_2{=}CH_2 + \ldots.$$

$$\text{—}CH_2\text{—}CH_2\text{—}CH_2\text{—}CH_2\text{—}CH_2\text{—}CH_2\text{— etc.}$$

polythene

Polythene has a molecular weight of about 30,000. It is an excellent insulator and is acid-resistant. It may be blown or cast into thin films and moulded to make various types of container. Polythene has been used as an underlay for the M1 motorway and in the construction of electric cables such as the transatlantic telephone cable.

Polyvinyl Chloride (P.V.C.)

Polyvinyl chloride is also made from ethene. The gas is first converted to vinyl chloride which is then heated with benzoyl peroxide. Benzoyl peroxide is only required in relatively small amounts and is technically known as an 'accelerator' since it speeds up polymerization.

$$\underset{\text{ethene}}{CH_2{=}CH_2} \xrightarrow{Cl_2} \underset{\text{1:2-dichloroethane}}{ClCH_2CH_2Cl} \xrightarrow{600°C} \underset{\text{vinyl chloride}}{CH_2{=}CHCl}$$

$$CH_2{=}CHCl + CH_2{=}CHCl + CH_2{=}CHCl + \ldots. \rightarrow$$

$$\ldots\text{—}CH_2\underset{}{\overset{\overset{Cl}{|}}{C}}H CH_2\overset{\overset{Cl}{|}}{C}H\cdot CH_2\overset{\overset{Cl}{|}}{C}H\text{—}\ldots\text{ etc.}$$

polyvinyl chloride

Like polythene it is an excellent electrical insulator and it shows good resistance to weather, corrosive chemicals and even fire.

Polystyrene

Polystyrene is derived from benzene and ethene. Benzene reacts with ethene in the presence of aluminium chloride to form ethyl benzene.

$$C_6H_6 + CH_2{=}CH_2 \xrightarrow{AlCl_3} C_6H_5C_2H_5$$

Ethyl benzene

Ethyl benzene is converted to styrene by heating to 550°C in the presence of zinc oxide.

$$C_6H_5C_2H_5 \xrightarrow[ZnO]{550°C} C_6H_5CH{=}CH_2 + H_2$$

Ethyl benzene → Styrene

Styrene is polymerized by refluxing with benzoyl peroxide.

$$n\ C_6H_5CH{=}CH_2 \longrightarrow \left[-CH(C_6H_5)-CH_2-CH(C_6H_5)-CH_2-CH(C_6H_5)-\right]_n$$

Styrene → Polystyrene

Polystyrene is used mainly in the electrical industry as an insulating material. It shows good resistance to corrosion and oxidation. It is glass-like in appearance and has been used to make plastic lenses.

Polyurethane

Polyurethane is produced by a more complicated sequence of reactions. It is made by polymerizing a diol with a di-*iso*cyanate.

$$n\ HO{\cdot}R'OH + n\ OCN{\cdot}R{\cdot}NCO \longrightarrow \cdots\left[O{\cdot}R'O{\cdot}OC{\cdot}NH{\cdot}R{\cdot}NH{\cdot}CO\right]_n\cdots$$

Diol, di-*iso*cyanate, Polyurethane

1:2-Dihydroxyethane is an example of a simple diol.

$$\begin{array}{c} CH_2OH \\ | \\ CH_2OH \end{array}$$

The di-*iso*cyanates are obtained from di-primary amines by reaction with carbonyl chloride.

$$NH_2R{\cdot}NH_2 + \underset{\text{carbonyl chloride}}{2COCl_2} \rightarrow ClCOHNRNHCOCl + 2HCl$$

$$ClCOHNRNHCOCl \xrightarrow{-2HCl} OCNRNCO$$

The polyurethanes are rubber-like plastics and are used for improving the abrasion resistance of tyre rubber. Polyurethanes are also used for making synthetic fibres and paints.

Nylon

Nylon is a polyamide made by polymerizing amino-acids or a mixture of diamines and dicarboxylic acids. The most important polyamide is Nylon 6.6 discovered in the U.S.A. in 1935 by Carothers.

Nylon 6.6 is made by heating the hexamethylene diamine salt of adipic acid.

$$n\ NH_2(CH_2)_6NH_2 + n\ COOH\cdot(CH_2)_4\cdot COOH \rightarrow \left[\bar{O}OC(CH_2)_4CO\bar{O}\ \overset{+}{N}H_3(CH_2)_6\overset{+}{N}H_3\right]_n$$

Hexamethylene diamine — Adipic acid — The salt

250 °C

$$\bar{O}OC(CH_2)_4CO\left[NH(CH_2)_6NH\cdot CO(CH_2)_4CO\right]_{n-1}NH(CH_2)_6\overset{+}{N}H_3$$

Nylon 6·6

The polymer has a molecular weight of about 10,000 and is referred to as Nylon 6·6 because it is derived from two substances each containing six carbon atoms. The world production of Nylon, used for making synthetic fibre, is at present about 3000 million pounds per annum.

Terylene

An American name for this substance is Dacron. It is a polyester and may be obtained from the reaction between methyl terephthalate and 1:2-dihydroxyethane.

$$nCH_3OCO\cdot C_6H_4\cdot COOCH_3 + n\begin{matrix}CH_2OH\\|\\CH_2OH\end{matrix} \rightarrow \left[CO\cdot C_6H_4\cdot CO\cdot OCH_2CH_2O\right]_n + 2nCH_3OH$$

Methyl terephthalate — 1:2 dihydroxyethane — Terylene

The raw materials required for its manufacture are derived from petroleum (see *Figure 11.1*). Terylene is used for making synthetic fibre. The present world production is about 1000 million pounds per annum.

Both Nylon and Terylene are treated in what is called a 'melt spinning' process to obtain fibre. The polymer is fed onto a melting grid. In the molten state it is forced through small holes in metal blocks called 'spinnerets'. On emerging from the spinnerets it is cooled by air and then wound onto large bobbins (*Figure 11.2*).

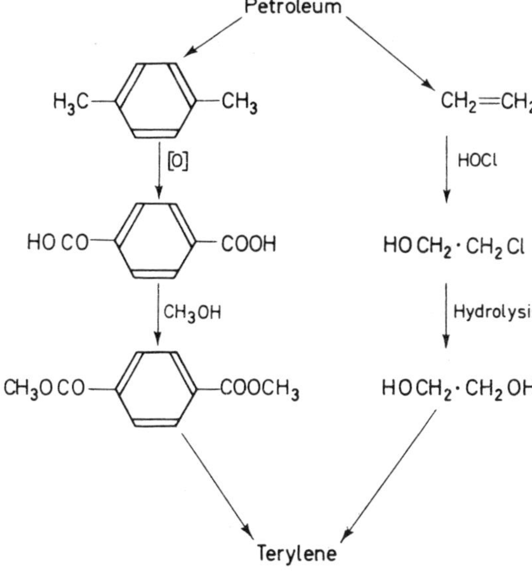

Figure 11.1

Semi-synthetic Fibres Derived from Cellulose

All natural vegetable fibres such as cotton, linen and hemp consist of cellulose and some important man-made fibres are derived from cellulose. The two most popular semi-synthetic cellulose fibres are Rayon and artificial silk (cellulose acetate). The world production of these is approaching 7000 million pounds per annum.

Rayon—This is made by treating wood pulp (impure cellulose) with carbon disulphide, which gives an orange-coloured powder that dissolves in 3% caustic soda solution to form a viscous liquid. This is known as 'viscose' and after being allowed to stand for a few days it is extruded from a spinneret into an acid solution in which it sets forming cellulose fibre.

$$ROH + CS_2 + NaOH \rightarrow [RO\overset{\overset{\displaystyle S}{\|}}{C}\text{—}\bar{S}]\overset{+}{N}a + H_2O$$

$$[RO\overset{\overset{\displaystyle S}{\|}}{C}\text{—}\bar{S}]\overset{+}{N}a + NaHSO_4 \rightarrow \underset{\text{cellulose}}{ROH} + CS_2 + Na_2SO_4$$

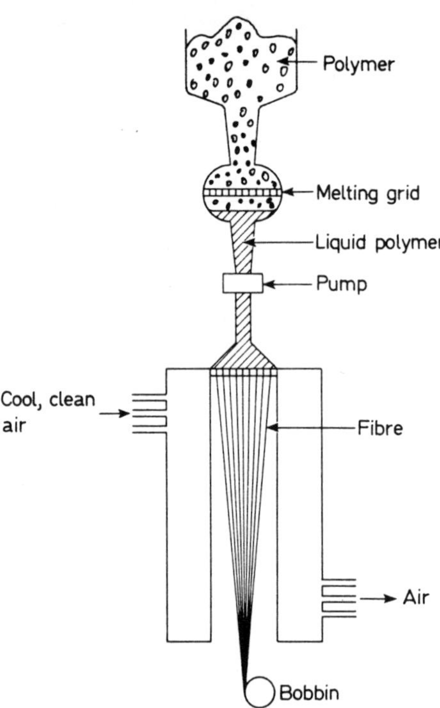

Figure 11.2. *Melt spinning apparatus*

Acetate Fibre—This is obtained by treating presoaked cellulose with acetic anhydride and partially hydrolysing the product with water to cellulose diacetate. The diacetate is dissolved in acetone or methylene chloride and the solution forced through spinnerets into warm air. The solvent evaporates leaving silk thread.

Thermosetting Plastics

In some polymerization processes, reaction occurs with the elimination of water (or some other simple substance). In addition, the polymer chains may be able to cross-link with one another forming a rigid structure. Under certain conditions formaldehyde and phenol react with one another in this way forming a hard, brown brittle resin known as Bakelite.

The physical properties of the thermosetting plastics can be adjusted by incorporating various 'fillers' such as asbestos and nylon.

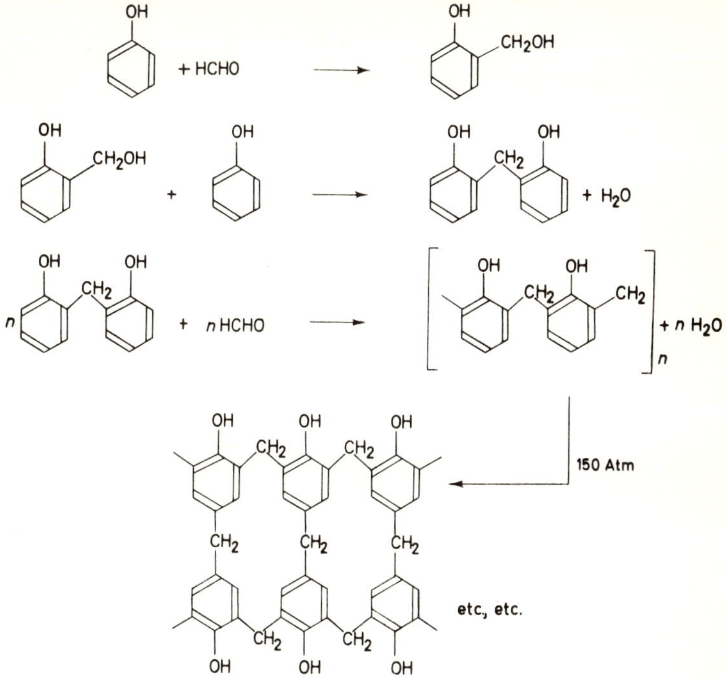

As a result of this, a variety of plastic materials can be made finding application in almost every industry..

In some cases phenol is replaced, with advantage, by urea or melamine to produce white plastics.

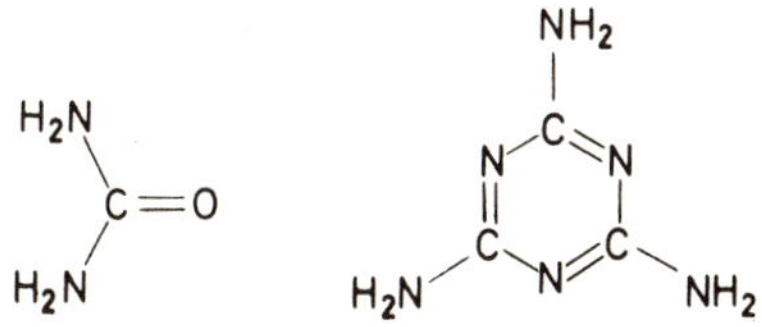

Urea Melamine

12

RUBBER

NATURAL RUBBER

Extraction and Treatment

RUBBER trees grow in many countries all round the world in a region just north and south of the equator. In cultivated plantations the rubber tree, *Hevea brasiliensis,* is the most common.

The trees are hardy and not difficult to cultivate. They grow in acid soil (pH = 5) which is fed with compound fertilizers containing nitrogen, phosphorus, potassium and magnesium.

The rubber is obtained from a milky fluid called 'latex' which is to be found in the phloem of the tree (*Figure 12.1*).

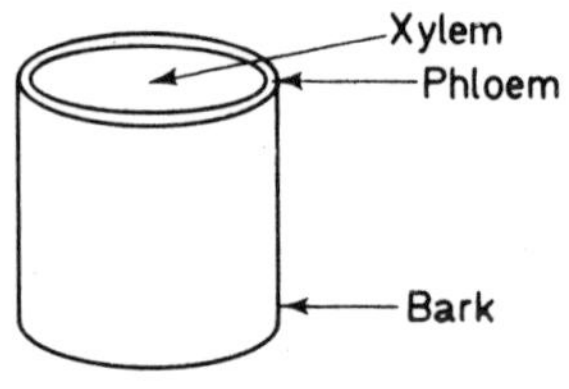

Figure 12.1

Latex is thought to be a waste product of the tree and is contained in latex vessels which originate from the cambium and move outwards into the phloem.

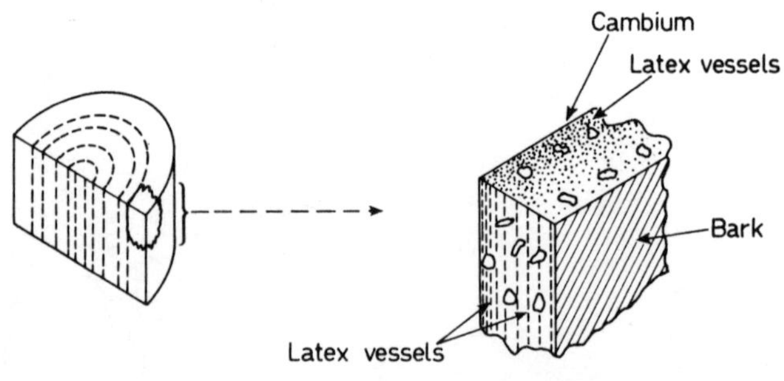

Figure 12.2

The latex is tapped from the tree by cutting away a small diagonal strip of bark about 1/8 in. thick. Latex weeps from the phloem, runs along the remaining bark and is collected (*Figure 12.3*).

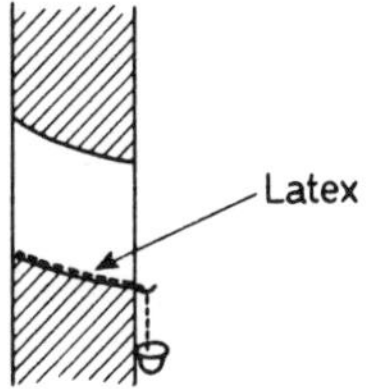

Figure 12.3

Latex is colloidal. It contains very small spherical particles of rubber (0·02 μ to 0·2 μ)* suspended in water. A dilute solution of latex shows the Tyndall effect and Brownian movement. The Tyndall effect may be observed as bright moving specks of light when a beam of light is focused into a colloidal solution and the solution is observed at right angles to the beam of light. It is caused by reflection from the moving colloidal particles. Brownian motion is the erratic movement of particles in a colloidal solution. It is a result of the bombardment of colloidal particles by solvent molecules. Both of these phenomena are peculiar to colloidal systems. (For a more detailed explanation of these phenomena the reader should refer to a physical chemistry textbook.)

The composition of latex is about 35% rubber, 3% other organic and inorganic substances and the rest water.

Treatment of Latex

After latex has been obtained from the tree it is treated to separate the rubber from the water. The method of treatment depends on the purpose for which the rubber is required. In one method the latex is centrifuged and the rubber concentrate, preserved with ammonia, is transported in tankers to the manufacturer. The alternative process involves coagulating the rubber by making the latex acid. The spongy rubber formed in this way is compressed to expel excess water and then rolled into sheets. The crude sheet rubber is cut and then dried and preserved by being smoked in large ovens. Smoking takes four to five days after which time the rubber is packed and transported to the manufacturer.

Treatment of Crude Rubber

Crude rubber, obtained by centrifuging latex, is used for making

*μ = 10^{-4} cm.

foam rubber (for upholstery and rubber toys) and also for making such things as rubber boots. Foam rubber is made by mixing the centrifuged latex to trap small air bubbles and then allowing it to set, having previously added certain gelling agents to the mix. Rubber boots can be made by dipping a mould into the crude rubber and then into a weak acid solution to cause coagulation and setting. By repeating this process a number of times the required thickness of rubber can be achieved.

Crude rubber in the form of sheet is used primarily to make heavy-duty rubber products such as tyres, mats and tiles. In the manufacture of tyres, the crude sheet rubber is transformed from a soft translucent plastic material to the relatively-hard, black, durable substance with which we are familiar. Tyre rubber has these characteristic properties as a result of a complex treatment involving 'plasticizing' and 'compounding'. In the plasticizing process crude rubber sheet is cut into small pieces and warmed to produce a spongy, doughy mass. When the rubber is in this state, various additions are made (compounding) to give the properties required in the final product. By adding about 2% sulphur, the rubber is made less plastic and more resilient. It is said to be 'vulcanized'. The process of vulcanizing may be accelerated by adding substances such as *cyclo*hexyl benzthiozyl sulphonamide called 'accelerators'. These may be likened to catalysts and are caused to function properly by substances known as 'activators'.

To harden the rubber, 'fillers' are added. Chalk and clay are examples of inexpensive fillers. Carbon black is added to rubber, used for tyres, to give added strength and resistance to wear.

Finally, substances referred to as 'antioxidants' are added to prevent oxidation and subsequent cracking. Anilines and phenyl β-naphthylamine are examples of antioxidants.

The Chemistry of Rubber

Rubber extracted from latex is a hydrocarbon (p. 3). Chemical analysis reveals that it contains approximately 88% carbon by weight, the rest being hydrogen. A rubber molecule is of very high molecular weight and comprises a large number of simple chemical units named 'isoprene units'. These have the formula and structure:

$$C_5H_8 \left\} \begin{array}{ccccccccc} & H & & CH_3 & & & & H & \\ & | & & | & & & & | & \\ - & C & - & C & = & C & - & C & - \\ & | & & & & | & & | & \\ & H & & & & H & & H & \end{array} \right. \qquad \text{Isoprene unit}$$

In a rubber molecule about 10,000 of these units are joined together end to end to form what is known chemically as polyisoprene. The rubber molecule is not perfectly straight. The reason for this is that the chemical bonds from each carbon atom are directed towards the corner of a regular tetrahedron and this results in the molecule having a zig-zag shape. Each kink in the chain represents a carbon atom.

Part of a rubber molecule; 50 carbon atoms represented

When raw rubber is heated, it softens and becomes very plastic. This happens because the long-chain molecules become, to a certain extent, disengaged and untangled thereby being able to move over one another more easily. Vulcanized rubber differs in that its structure is much more rigid. When sulphur is used as the vulcanizing agent, it acts as a bridge between rubber molecules preventing easy movement over one another.

```
            CH3                 CH3                  CH3
            |                   |                    |
—CH2┼CH2C=CH·CH2┼CH2C=CH·CH2┼CH2C=CH·CH2┼CH2—
                                                          Parts of two
                                                          rubber molecules
—CH2┼CH2C=CH·CH2┼CH2C=CH·CH2┼CH2C=CH·CH2┼CH2—
            |                   |                    |
            CH3                 CH3                  CH3
```

Unlinked chains

```
            CH3                 CH3                  CH3
            |                   |                    |
—CH2┼CH2C—CH·CH2┼CH2C—CH·CH2┼CH2C—CH·CH2┼CH2—
            |     |             |     |              |     |
            Sn    Sn            Sn    Sn             Sn    Sn      Parts of two
            |     |             |     |              |     |       rubber molecules
—CH2┼CH2C—CH·CH2┼CH2C—CH·CH2┼CH2C—CH·CH2┼CH2—        held together by
            |                   |                    |             sulphur atoms
            CH3                 CH3                  CH3
```

Linked chains (vulcanized)

S_n = a small number of sulphur atoms bonded to one another and bridging two rubber molecules.

SYNTHETIC RUBBER

Synthetic rubber is a term applied to man-made polymers having properties similar to those of the natural product. Up to the present time, a synthetic rubber identical to that of natural rubber has not been made. It would appear that the chemical nature of natural rubber, described in the previous paragraph, is over-simplified since the substance polyisoprene has been synthesized and although the product is very similar, it is not *identical* to natural rubber. It seems probable that, in the foreseeable future, natural rubber will be exactly replicated but it will be some time before the product is manufactured economically.

At the present time, the most important and successful synthetic rubbers include S.B.R. (or Buna S), Butyl, Buna N and Neoprene.

S.B.R. (Styrene–Butadiene Rubber)

This is made by polymerizing styrene with butadiene in the presence of catalysts.

$$n\,CH_2{=}CH{\cdot}CH{=}CH_2 + n\,CH_2{=}CH{\cdot}C_6H_5 \rightarrow \left[-CH_2{\cdot}CH{=}CH{\cdot}CH_2\,CH_2\underset{\displaystyle C_6H_5}{\underset{|}{C}H}-\right]_n$$

S.B.R.

In the modern process styrene and liquid butadiene are emulsified in water. The monomers react together at low temperature, in the presence of various catalysts, to form the final co-polymer.

S.B.R. is usually compounded with carbon black and vulcanized with sulphur. It is very similar to natural rubber and is used for making tyres.

Butyl Rubber

This is another co-polymer rubber made by combining *iso*butylene (2-methylprop-1-ene) and isoprene (2-methylbuta-1:3-diene).

$$60\,n\;\overset{\displaystyle CH_3}{\overset{|}{\underset{\displaystyle CH_3}{\underset{|}{C}}}}{=}CH_2 \;+\; n\,CH_2{=}\overset{\displaystyle CH_3}{\overset{|}{C}}-CH{=}CH_2 \rightarrow \left[\left(-\overset{\displaystyle CH_3}{\overset{|}{\underset{\displaystyle CH_3}{\underset{|}{C}}}}-CH_2-\right)_{60} CH_2\overset{\displaystyle CH_3}{\overset{|}{C}}{=}CH{\cdot}CH_2-\right]_n$$

*iso*butylene — Butyl rubber

In the preparation of the polymer, 96% *iso*butylene is mixed with 4% isoprene. The mixture is diluted with methyl chloride and cooled to about $-100°C$. The polymerization is initiated by boron trifluoride in the presence of a minute amount of water.

The final product is unusual in that it is very impermeable to air and resistant to oxidation by oxygen and ozone. It is used for making tyre inner tubes, steam hosing and gaskets.

Buna N (Butadiene–Acrylonitrile Rubber)

This rubber is made by polymerizing butadiene and acrylonitrile (vinyl cyanide).

$$2n\ CH_2{=}CH{\cdot}CH{=}CH_2 \quad + \quad nCH_2{=}\underset{\displaystyle CN}{\underset{|}{CH}} \rightarrow \left[-\!\!\left(CH_2CH{=}CH{-}CH_2 \right)_2 {-} CH_2\underset{\displaystyle CN}{\underset{|}{CH}}{-} \right]_n$$

Butadiene — Acrylonitrile — Buna N

Although Buna N is not produced in as large a quantity as S.B.R. and butyl rubber, it is useful since it shows remarkable resistance to oils and solvents. For this reason, it is used for making petrol hoses, solvent containers, gaskets and chemical plant.

Neoprene

This is made from a single monomer, chloroprene (2-chlorobuta-1:3-diene).

$$CH_2{=}\overset{\displaystyle Cl}{\overset{|}{C}}{-}C{=}CH_2$$

The substance is emulsified in water using soap. This emulsion is then treated in the usual way with initiator and modifiers to obtain chains of the required length, and 'stoppers' to terminate the reaction.

Neoprene resembles smoked rubber in appearance. In some respects neoprene is better than the natural product; it has, for example, a high resistance to solvents, heat, light and ozone. It is used for making transmission belts, hose covers, printing rollers and flooring materials.

SUMMARY

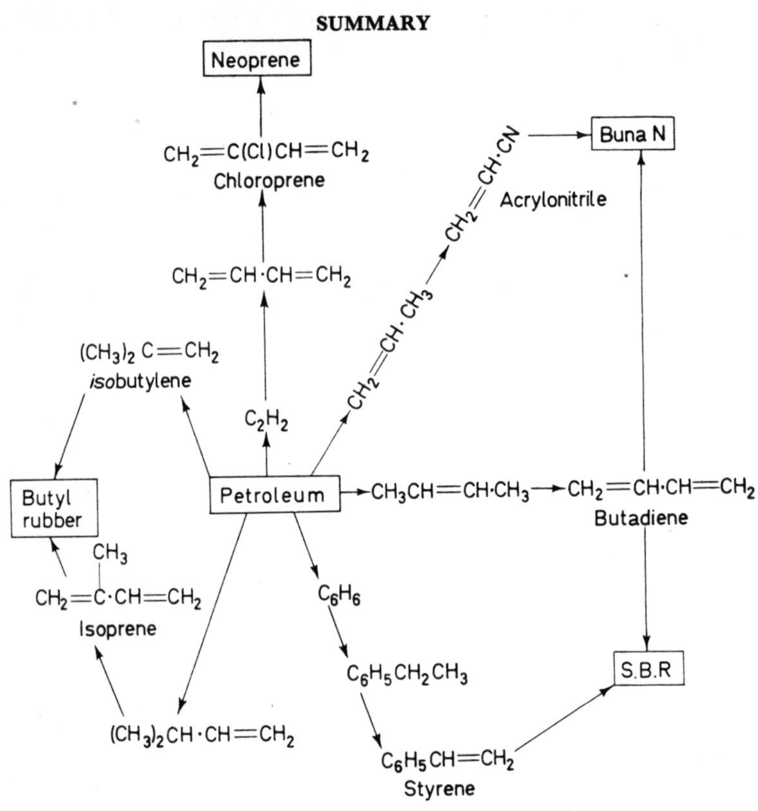

Figure 12.4

13

DYES

DYES, for colouring cloth, have been used for thousands of years but it is only relatively recently that man has discovered how to synthesize dyes from readily-available substances. In the past they were obtained from plants and rare sea creatures.

ALIZARIN AND INDIGO

Two of the best-known dyes are alizarin (red) and indigo (blue). Alizarin is a mordant dye, so called because it has little affinity for cloth and therefore must be held onto the cloth by a substance called a mordant. The most commonly used mordants are basic oxides, e.g. aluminium, tin and chromium oxides.

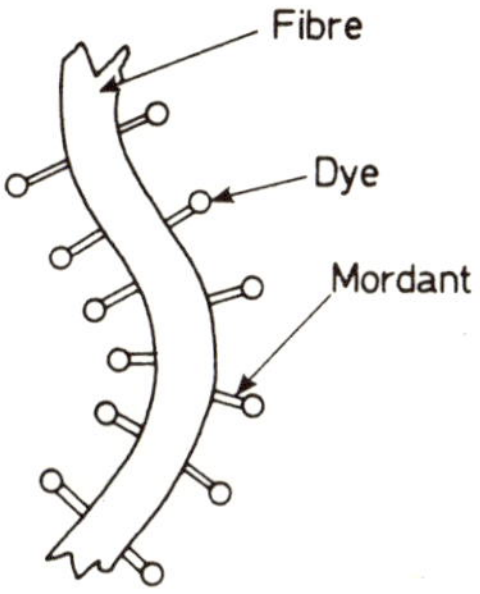

Pure alizarin forms beautiful red crystals which are sparingly soluble in water. With mordants the dye forms insoluble complexes called 'lakes' and the colour of these depends on the metal compound used as mordant. With aluminium oxide the dye forms red lakes, and with chromium oxides it forms violet-coloured lakes.

Alizarin was originally obtained from the madder plant. This grows to about three feet in height and prior to about 1870 was specially cultivated in France and Britain for producing the dye. It is interesting to note that before 1870 alizarin, derived from the madder plant, was used to dye the uniforms of British and French soldiers; this was enforced by royal decree in order to safeguard agricultural interests.

In 1865 two German chemists, Graebe and Liebermann, discovered that by distilling alizarin with zinc dust a hydrocarbon, known as anthracene, was obtained. This, and further experiments, led them to the conclusion that the alizarin molecule must contain a nucleus very similar to anthracene. The molecule was in fact found to be a dihydroxyanthraquinone.

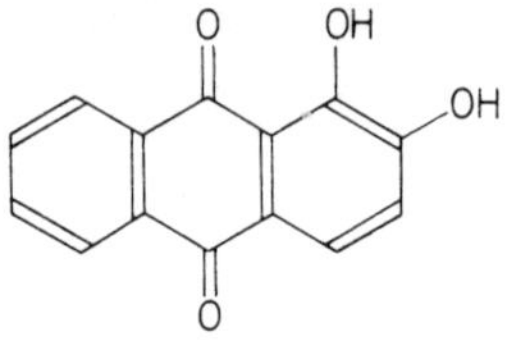

Alizarin (1:2-dihydroxyanthraquinone)

Compare this with anthracene and anthraquinone.

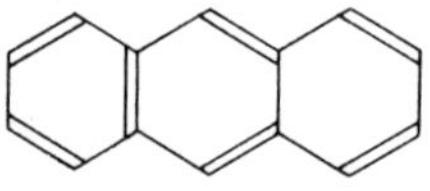

Anthracene

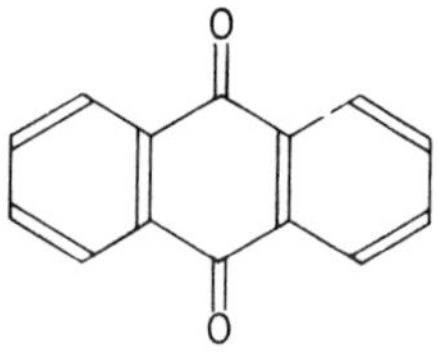

Anthraquinone

Graebe and Liebermann first made alizarin by fusing together dibromoanthraquinone and potassium hydroxide but this method was not adopted for the manufacture of alizarin because it was not economic (i.e. no cheaper than extracting the dye from the madder plant). It was not until 1865, that patents were filed for the manufacture of alizarin using the cheaper anthraquinone sulphonic acids.

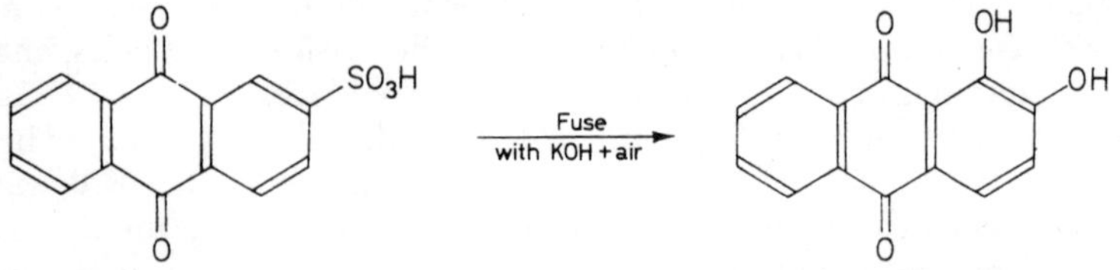

2-anthraquinone sulphonic acid $\xrightarrow[\text{with KOH + air}]{\text{Fuse}}$ Alizarin

It should be noted that although alizarin is well known, it is now used very little. It has been replaced by more effective anthraquinone dyes which are generally more complex than alizarin but they all contain the anthraquinone nucleus.

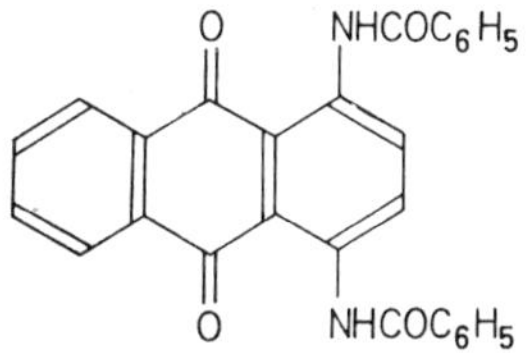

Indrathrene red 5GK

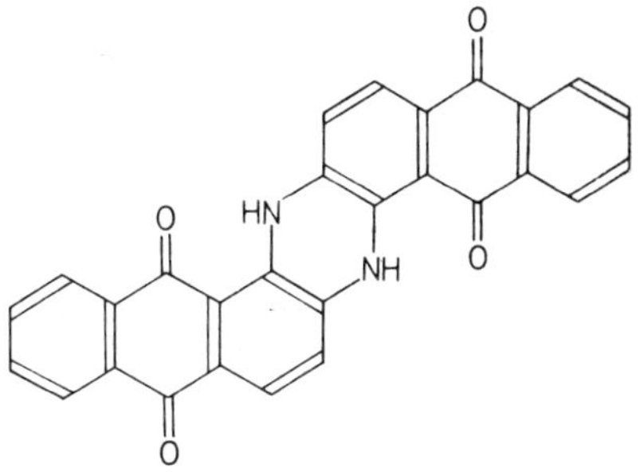

Indrathrene blue R

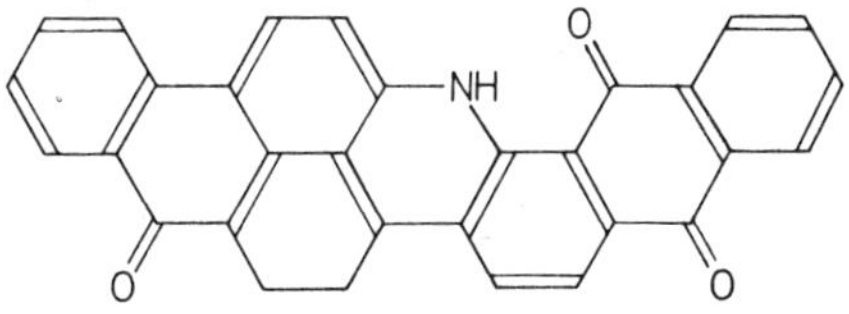

Vat green 3

Indigo dyes have been used for thousands of years. It is known, for example, that Egyptian mummy cloths, estimated to be over 4000 years old, were coloured with indigo. The dye was originally obtained from the woad plant and prior to 1900 the plant was cultivated in many parts of the world. Indigo was first synthesized by Baeyer in 1879 but it was not until 1897 that a method of preparation was found which was suitable for the manufacture of indigo, making the synthetic product commercially competitive with the natural product.

The raw material for the manufacture of indigo is naphthalene. This is converted to phthalic anhydride by the action of concentrated sulphuric acid in the presence of a mercury salt which acts catalytically. The anhydride is then treated with ammonia to form phthalimide and the latter is converted to anthranilic acid.

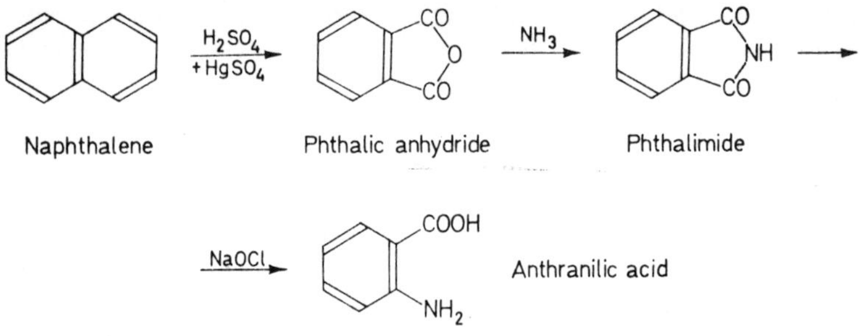

H_2SO_4 + $HgSO_4$ → ; NH_3 →

Naphthalene Phthalic anhydride Phthalimide

NaOCl → Anthranilic acid

Anthranilic acid is further treated with monochloroacetic acid to give *N*-phenylglycine-*O*-carboxylic acid.

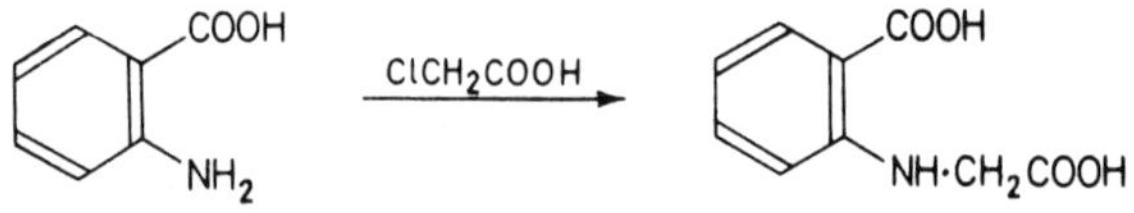

$ClCH_2COOH$ →

The *N*-phenylglycine is finally fused with sodium hydroxide and the product air-oxidized to indigo.

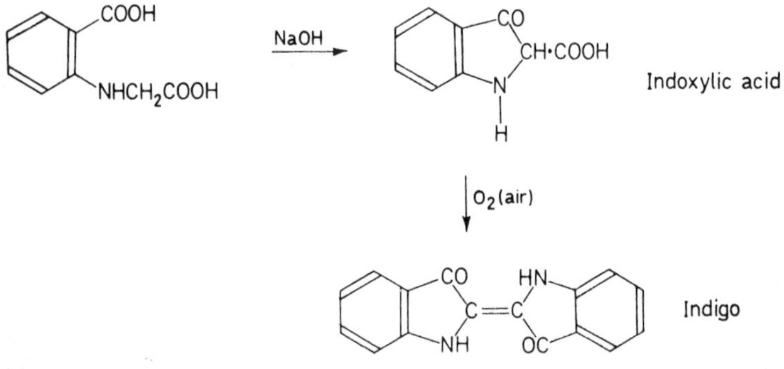

NaOH → Indoxylic acid

O_2 (air)

Indigo

Indigo is insoluble in water and therefore is not suitable for direct use as a cloth dye. It must first be reduced to a water-soluble form known as white indigo. The cloth is soaked in water-soluble white indigo and then the white form is oxidized, by air, back to the blue form.

Dyes like indigo, which form water-soluble reduced forms, are called vat dyes since the reduction and dyeing was at one time carried out in large open vats. Indigo dye is fairly fast (permanent) to light and washing and is a cheap dye to manufacture. For these reasons the dye is still used but in recent years its production has tended to fall off.

AZO DYES

The dyes manufactured today in largest quantity are the azo dyes. These are named, azo, diazo and triazo depending on the number of azo groups (—N=N—) in one molecule of the dye.

The azo dyes were discovered by Griess in 1862 and they were made by treating a phenol, or aromatic amine, with a substance known as a diazonium compound. Diazonium compounds contain the all important —N=N— group which gives the dye its colour (chromophore group).

The diazonium compound is prepared by treating an aromatic amine salt with nitr*ous* acid at low temperature (below 5°C).

$$C_6H_5\overset{+}{N}H_3\overset{-}{Cl} + HNO_2 \longrightarrow [C_6H_5N{=}N]^{+}Cl^{-} + 2H_2O$$

Aniline hydrochloride; Nitrous acid (obtained from $HCl + NaNO_2$); Benzene diazonium chloride

The diazonium salt is unstable and is not isolated. For the purpose of making a dye, a solution of a diazonium salt is added to a solution of phenol in alkaline solution.

$$[C_6H_5N{=}N]^{+}Cl^{-} + C_6H_5OH \xrightarrow[\text{alkaline}]{5°C} C_6H_5{-}N{=}N{-}C_6H_4{-}OH + HCl$$

Benzene diazonium chloride; Phenol; *p*-hydroxyazobenzene (yellow)

Although *p*-hydroxyazobenzene is coloured and will dye cloth, it is not used commercially. The azo dyes of commercial importance, although prepared by methods similar to that described for *p*-hydroxyazobenzene, are generally chemically more complex. The azo dye made in largest quantity is 'Direct Black 38' and this has the formula:

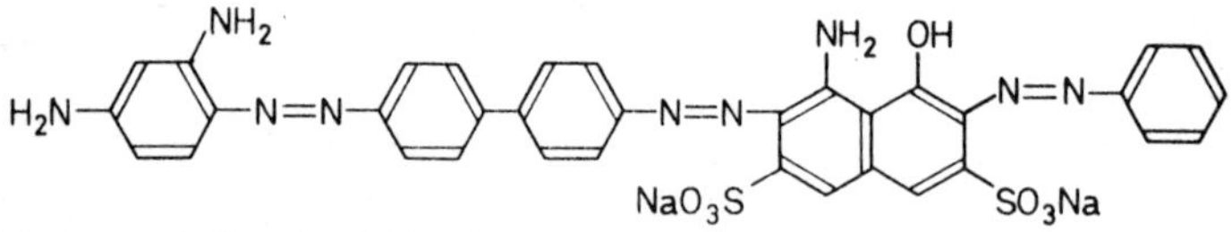
NH2
H2N
N=N
NH2 OH
N=N
NaO3S
SO3Na

It is made by the following sequence of reactions:

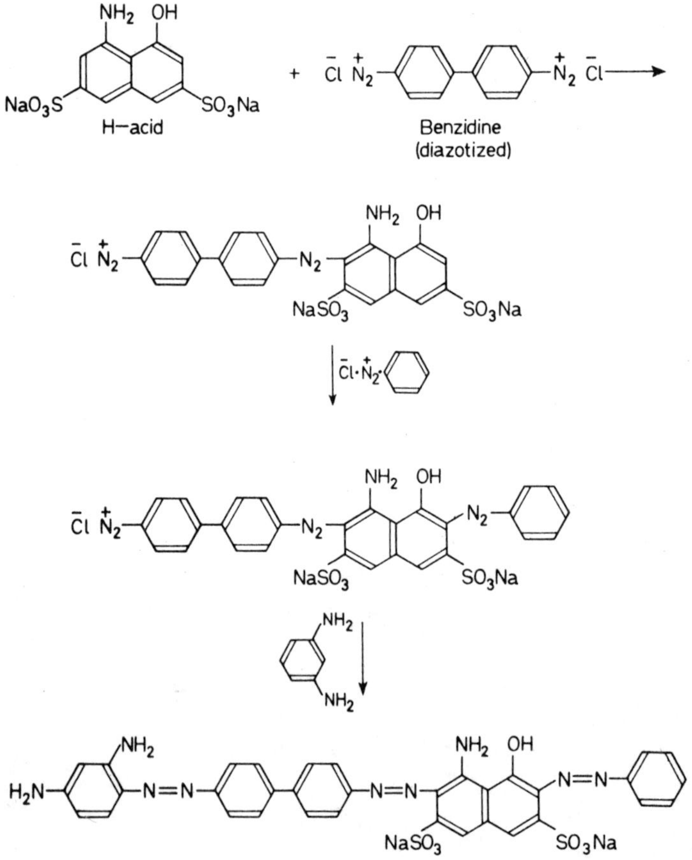
H—acid
Benzidine
(diazotized)

'Direct Black 38'

Many azo dyes are applied directly to the article to be coloured. They are used extensively to dye fibre and this may be effected simply by dipping into a suitable solution of the dye. Alternatively, the dye may actually be prepared on the fibre. This involves preparing solutions of the components of the dye and successively dipping the fibre in these solutions; cotton fibres may be dyed in this manner. A better method of dyeing cellulose fibre is by using mordants to fix the azo dye on the fibre.

The newer fibres, such as Nylon and Terylene, are dyed by making dispersions of the dyes in suitable solvents. Sometimes dyeing may be achieved by dissolving the azo dye, or modified azo dye, in a solvent in which the fibre is also soluble. By this means the fibre may be dyed before spinning. As an example, modified azo dyes, soluble in acetone, are used to dye cellulose acetate prior to spinning.

14

DRUGS

INTRODUCTION

THE term, drugs, refers to a whole series of substances which are in some way, directly or indirectly, concerned with our good health. Most of these substances are described in the British Pharmacopoeia but in this brief account only a few of the more important drugs will be considered. For our purpose we may classify drugs into two groups: (1) those which are soothing and/or pain killing (often affecting the autonomic nervous system), and (2) those which are chemotherapeutic, i.e. control parasitic infection in man by direct chemical action with the parasite. Included in the first group are cocaine, morphine, aspirin, phenacetin and the general anaesthetics, ether, chloroform, *cyclo*propane and halothane. Belonging to the second group are quinine, the sulphonamides and the antibiotics such as penicillin, streptomycin and chloromycetin.

GROUP 1

In the event of an illness, we are usually warned of what is to come by our very sensitive nervous system. For example, if we get a sore throat we take precautions against a cold; if we have an aching tooth we visit the dentist; and if we strain a muscle we rest until the muscle has recovered because we know from experience that the muscle will continue to be painful until it has recovered. Sometimes the body seems to warn us unmercifully even after we have acknowledged that something is amiss and at times like this, we can stop or subdue the pain by taking drugs. In the case of a headache or slight toothache, a mild pain reliever (analgesic) such as aspirin is used, but if the pain is more severe, as for example in a surgical operation, then a more powerful drug is required (an anaesthetic) such as *cyclo*propane, Novocaine or morphine.

Aspirin

Aspirin is an antipyretic and analgesic (soothes fever and alleviates pain). It is widely used and has replaced salicylic acid from which it is derived. Salicylic acid was the first antipyretic but was not ideal as such because it produces undesirable side effects such as tissue irritation. Aspirin can be made from salicylic acid (a constituent of Oil of Wintergreen) by treating the acid with acetyl chloride (acetylation).

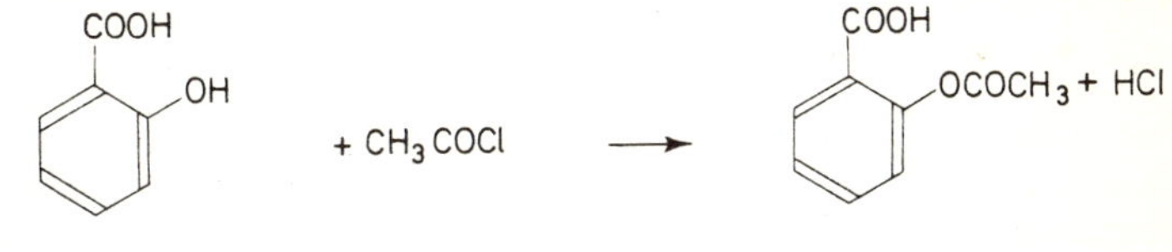

$$C_6H_4(COOH)(OH) + CH_3COCl \longrightarrow C_6H_4(COOH)(OCOCH_3) + HCl$$

Salicylic acid — Acetyl chloride — Aspirin (acetyl salicylic acid)

Aspirin is a white crystalline solid (m.p. 135°C), sparingly soluble in water.

Cocaine

Cocaine is a white crystalline solid (m.p. 98°C) which may be extracted from the leaves of the coca plant (not to be confused with the cocoa tree, used to make chocolate). The drug was first isolated by Niemann in 1860 and first used as a local anaesthetic by Koller and Freud in 1884. Cocaine, at one time widely used as a local anaesthetic, has now a rather restricted use in eye operations. The disadvantages of cocaine are that it is toxic and when taken in large quantities, or regularly, is addictive. For these reasons alternatives to cocaine had to be found and this involved a full investigation of cocaine itself. The procedures used for determining the structure of cocaine are outlined in Chapter 9. Results of analyses show that cocaine has the structure:

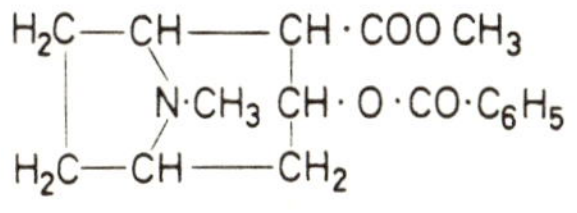

$H_2C—CH——CH \cdot COO\,CH_3$

$N\cdot CH_3 \quad CH \cdot O \cdot CO \cdot C_6H_5$

$H_2C—CH——CH_2$

Cocaine

Once this structure was known, substances having similar chemical structures were synthesized and their therapeutic action investigated. Of these substances Novocaine is probably the best known. It is still widely used because, although more powerful compounds have since been discovered, it is easy to handle and not as toxic as, for example, cocaine.

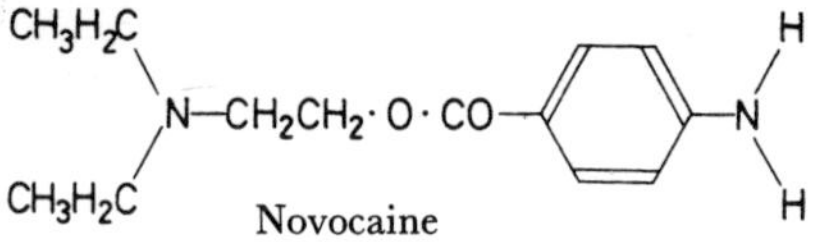

Novocaine

GROUP 2

Many illnesses are caused by the presence of parasites in our bodies. These may be large creatures like liver flukes or tape-worms, or they may be microscopic organisms such as protozoa, bacteria and viruses. The drugs included in this group are used to effect a cure of such illnesses by making contact with, and destroying the parasites. Generally speaking, the smaller the parasite the more difficult it is to detect and destroy and today a good deal of research is being done to find substances which are active against common virus diseases. The common cold is an example of a virus disease for which there is still no cure. This is not altogether surprising when we realize that virus parasites are organisms nearly a million times smaller than the common house-fly.

Some of the more familiar chemotherapeutic drugs will now be considered.

Antimalarials

The disease malaria is caused by protozoa originating from the female *Anopheles* mosquito. It is still very common and widespread but it can now be controlled providing the drugs are available in sufficient quantity. At one time, quinine was the only substance known to be active against the disease and although we now have more effective antimalarial drugs, quinine is still used in countries where the disease is common.

The structural formula of quinine is:

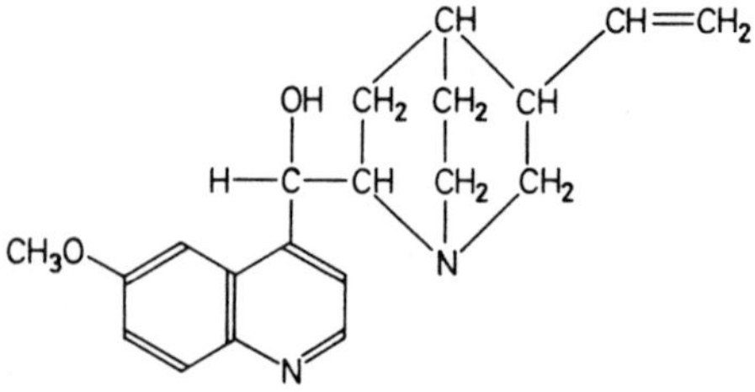

Quinine

In 1944 (the year that quinine was synthesized) a new anti-malarial drug was produced by I.C.I. This was called paludrine (paludal: of a marsh) and it has been found to be even more effective against malaria than quinine. This is rather surprising since its chemical formula in no way resembles that of quinine.

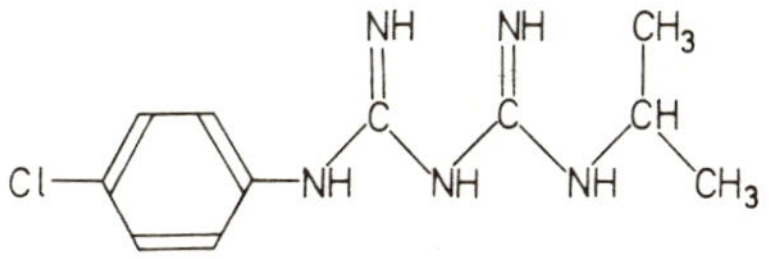

Paludrine

Sulphonamides

Paul Ehrlich, in 1887, discovered that a number of dyes only coloured certain animal tissue. He found, for example, that the dye methylene blue was specific for the cells of the nervous system. This discovery suggested to Ehrlich that it might be possible using certain dyes to attack and destroy the micro-organisms causing many illnesses. After many experiments on mice he found an arsenical substance (known as 'salvarsan') to be effective against trypanosomes (protozoa) which cause a fatal illness in mice. The salvarsan eliminated the trypanosomes and was harmless to the mice. When salvarsan was tried on humans, it was unfortunately found to be without action on the trypanosomes causing sleeping sickness. Surprisingly, it did kill some micro-organisms which are the cause of syphilis, namely *Spirochaeta pallida*, which are bacteria.

Following the work of Ehrlich, chemists all over the world investigated the action of thousands of dye-like substances in an effort to find substances effective against the many diseases which were, at that time, fatal to man. As a result of these efforts, a number of chemicals were found to be useful. The azo dye 'prontosil' was found to be active against the bacteria which cause scarlet fever. It has a chemical structure very similar to that of 'salvarsan'.

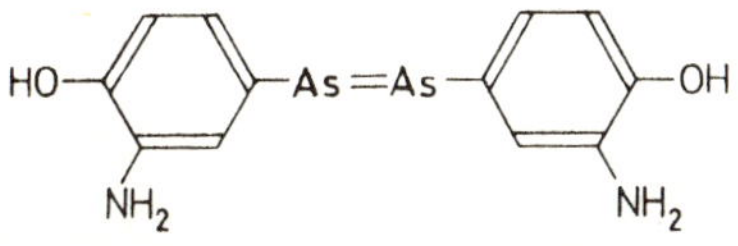

Salvarsan

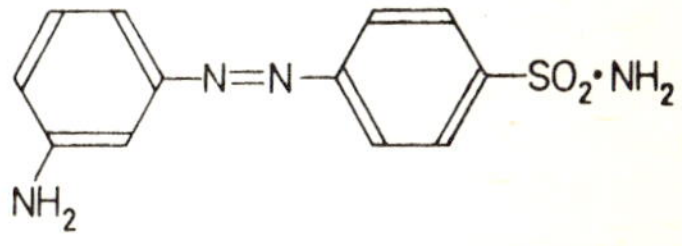

Prontosil

Prontosil is a sulphonamide because it contains the —SO_2NH_2 group.

In 1938 May & Baker (a British Pharmaceutical firm) working on sulphonamides produced M & B 693. This substance has the structural formula:

NH_2—C_6H_4—SO_2·NH—C_5H_4N

M & B 693

(*p*-aminobenzenesulphonamide-2-pyridine)

This substance was important because it is effective against bacteria causing pneumonia and meningitis. Later M & B 693 was replaced by M & B 760 which has the structural formula:

NH_2—C_6H_4—SO_2·NH—C_3H_2NS

M & B 760

(*p*-aminobenzenesulphonamide-2-thiazole)

Other useful sulphonamide drugs have been made by replacing pyridine and thiazole in M & B 693 and 760 by other nitrogen-containing groups.

Research into the action of the sulphonamide drugs has shown that the drug does not kill the bacteria by direct contact. The drug competes with certain metabolites of the bacteria for vital enzyme systems essential for the reproduction of the bacteria.

Antibiotics

Antibiotics are substances which are produced in natural processes and which inhibit the reproduction of certain micro-organisms. Penicillin was the first antibiotic to be discovered. In 1928, Sir Alexander Fleming noticed that some of his bacterial cultures failed to grow when contaminated with a certain airborne mould. When the mould was examined, it was found to be an uncommon variety known as *Penicillium notatum*. Later work on this mould showed that the chemical substance (penicillin) produced by the mould and active against the bacteria, was only produced in small quantity. Penicillium notatum was not therefore suitable as a commercial source of the drug penicillin. An intensive search resulted in the discovery of *Penicillium chryogenum* which produced penicillin in sufficient quantity for it to be used as a source of the drug.

Since the first discovery made by Sir Alexander Fleming, many penicillins have been cultured and synthesized. They have the general formula:

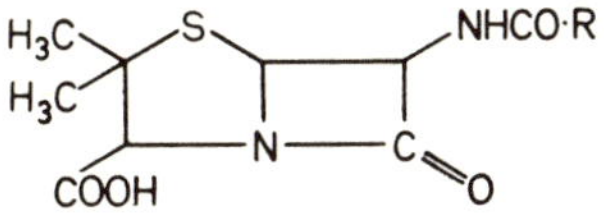

If R = 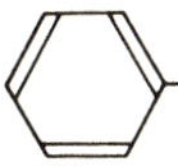 $-CH_2-$ (benzyl group) the penicillin is known as 'penicillin G'.

Some antibiotics such as streptomycin attack bacteria which are otherwise immune. Streptomycin can be used to treat tuberculosis. The bacteria causing this disease are protected by a fat layer which can only be penetrated by antibiotics like streptomycin.

Chloromycetin and terramycin are antibiotics which are active against some viruses. Viruses live in the cells of tissues, and diseases caused by viruses are extremely difficult to treat. The problem is to find a substance which is active against the virus but which does not affect the cell metabolism. Chloromycetin and terramycin are two chemicals which are successful in this respect.

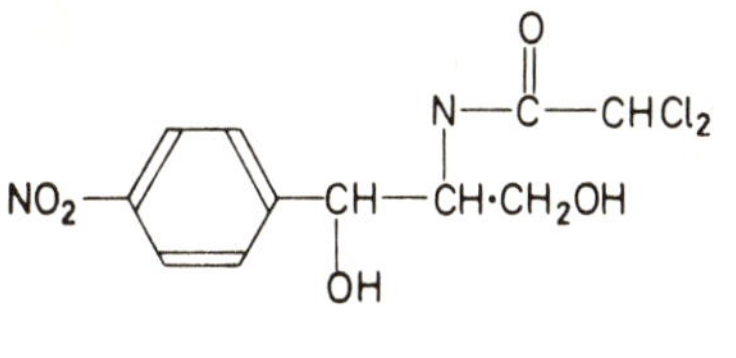

Chloromycetin

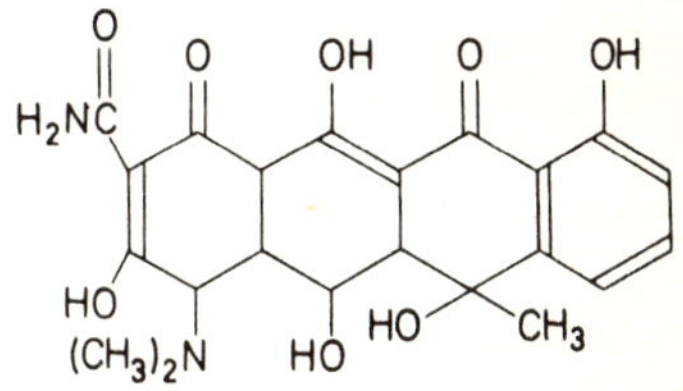

Terramycin

INDEX